Field engineering for
agricultural development

Field engineering for agricultural development

N. W. HUDSON

CLARENDON PRESS · OXFORD

1975

Oxford University Press, Ely House, London W. 1

GLASGOW NEW YORK TORONTO MELBOURNE WELLINGTON
CAPE TOWN IBADAN NAIROBI DAR ES SALAAM LUSAKA ADDIS ABABA
DELHI BOMBAY CALCUTTA MADRAS KARACHI LAHORE DACCA
KUALA LUMPUR SINGAPORE HONG KONG TOKYO

Casebound ISBN 0 19 859442 9
Paperback ISBN 0 19 859456 9

Reproduced and printed by photolithography and bound in
Great Britain at The Pitman Press, Bath

Preface

Agricultural engineering is important in almost all developing countries. There are few countries where the food requirement can be met solely by improving traditional agriculture. Most countries have to think in terms of opening up new land and improving the management of the land already used. The technique of developing, managing, and improving soil and water resources is what this book is about. It does not try to cover the whole range of agricultural engineering, for it does not include mechanization, buildings, or crop-handling. It covers only that part of agricultural engineering which in the United States is called 'soil and water engineering', and in Britain 'field engineering'.

The purpose of the book is to help bridge the gap between the unskilled layman and the professional agricultural engineer. Ideally agricultural development should be planned and carried out by experienced specialists, but most developing countries are still building up their numbers of trained men. So agricultural development today involves everyone who can help, whether they are agricultural officers, veterinary officers, foresters, extension workers, or local government officials. This book aims to provide how-to-do-it information for people like this. It does not cover the whole subject, it takes short-cuts, it oversimplifies, and it often neglects to explain why one particular method is recommended. My justification for all these failings is that I think the book will be most useful if its main object is to be simple and practical.

I have avoided all mathematics except simple arithmetic, but I have assumed that anyone concerned with this kind of work will have a basic knowledge of surveying. The subject matter is not directed to any examination syllabus, but would be suitable for Diploma-level teaching in agricultural engineering, or as a subsidiary subject in a degree course in agriculture, geography, or forestry.

Bedford, England
July 1974
N.W.H.

Contents

A note on units

It is not easy to make this kind of book equally usable in both English and metric units. The dimension of a pipe may be converted from one unit to the other, but this is not meaningful since the pipe will be made to either a standard size in inches or a standard metric size. Also, empirical design procedures which have been developed in one set of units do not convert readily.

Since the world trend is towards metrication, the international metric system (SI) has been used. The main differences between SI and technical metric units are as follows.

(1)   SI uses a primary unit (for example, the metre) with multiples $\times\ 10^3$ (the kilometre) and subdivisions $\times\ 10^{-3}$ (the millimetre). Multiples of 10^2 and subdivisions of 10^{-2} are not used in SI, so the centimetre disappears, and dimensions are given only in metres (m) and millimetres (mm). There are a few exceptions allowed, and the hectare (1 ha $= 10^4$ m^2) is one of them.

(2)   The SI unit of pressure is different from the commonly used units in both the English and metric systems.

Data required for design purposes has been also included in fps units, but no attempt has been made to give every simple dimension in both units. When changing a design procedure from one set of units to another I have used equivalent practical values not a mathematical conversion. For example, the standard depths used in soil survey work of 10, 20, and 36 in have been replaced by 250, 500, and 1000 mm. A drain depth of 1 m has been translated to the equivalent practical unit of 3 ft rather than 39·37 in.

Several of the water measuring devices, such as the Washington flume, Parshall flume, and H flume, were originally designed in the United States and so built and calibrated in English units. The metric calibration figures given in the book are calculated conversions of the English figures because direct calibration has, to the best of my knowledge, not yet been carried out in metric units. Similarly some of the empirical design procedures, such as the Durbach method for stormwater drains, and the estimation of surface run-off, are metric conversions of procedures developed in English units.

Scales for conversion of some comman units are printed on p. 226, and other conversion factors are given in the Appendix.

1. Land selection

1.1. Soil survey

Selecting land for development means making choices and taking
decisions. To make the right decisions we need the relevant facts. The
purpose of soil surveys is to help to make the facts available. For
planning agricultural development it is necessary to have other information as well, for instance climatic and hydrographic data, so we often
refer to 'natural resource surveys' or 'land and water resource
inventories'. However the soil survey is usually the central feature. The
first step is to collect the facts—this is the survey. The second step is to
sort out the large amount of information into usable packages, and this
is the process of land classification.

The collection of data in the soil survey may be a general inventory
of all the data which might at some time in the future be useful, or it
may be done with a specific purpose in mind. In that case it will concentrate on the facts required for that purpose.

1.1.1. Kinds of soil surveys

The amount of detail in the collected information will depend upon
both the area to be surveyed and the scale at which the results are to be
mapped.

The range of different kinds of soil surveys can be illustrated by
some surveys in Africa. At one end of the scale is a survey of the whole
continent south of the Sahara desert. The purpose is to show
pedological soil units (that is, based on their origin and formation), and
the results are mapped at 1/5 000 000.

At the country level, the soil resources of Lesotho were mapped at
1/250 000 and those of Nigeria at 1/1 000 000. The purpose is to provide
an inventory of the total soil resources of the country, to assist the
selection of areas for more detailed study.

At regional level a feasibility study of irrigation development in parts
of Lesotho called for a reconnaissance survey at 1/50 000 with detailed
surveys at the same scale on those areas particularly likely to be chosen.

At project level detailed surveys might be produced down to 1/2500
for special purposes such as planning experimental plots.

Table 1.1
Soil Surveys

Descriptive name	Scale	Ratio of ground area to map area 100 mm²	Ratio of ground area to map area 1 in²	Approximate density of sampling points	Purpose
Reconnaissance	1 : 500000 to 1 : 250000	25 km² to 6·25 km²	62·3 mile² to 15·6 mile²	1 per 50 km² or 1 per 20 mile² to 1 per 10 km² or 1 per 5 mile²	Development planning at national scale
Detailed Reconnaissance	1 : 100000	1·0 km²	2·5 mile²	1 per 2 km² or 1 per mile²	Regional planning
Semi-detailed	1 : 50000	25 ha	400 acres	1 per 25 ha or 1 per 50 acres	Local planning; large agricultural projects, for example, irrigation
Detailed	1 : 25000	6·25 ha	100 acres	1 per 5 ha or 1 per 10 acres	Farm planning, irrigation or drainage projects
Intensive	1 : 10000 to 1 : 2500	1 ha to 0·06 ha	16 acres to 1 acre	1 per 2 ha or 1 per 5 acres to 1 per 0·5 ha or 1 per acre	Detailed farm planning; small projects, urban planning. Design of engineering works, for example, reservoirs, design of agricultural experiments.

The names 'reconnaissance', 'semi-detailed', etc. define different kinds of soil survey as shown in Table 1.1.

A *detailed survey* is one in which the location of the boundaries between different kinds of soil are observed all along their length.

A *reconnaissance survey* is less precise, and the boundaries are located by interpolation between inspected points set out on a rectangular grid.

A *semi-detailed survey* is one which has a special purpose and is therefore selective. A closer look is taken at the interesting parts than the rest. For example, in a survey of irrigation potential, the steep unirrigable land might be ignored, but a detailed study made of all the parts which might be irrigable.

Sampling. The most important thing about any sample is that it must be representative of whatever is being sampled, and this is true for soil sampling. If we had a soil map of the area it would be easy to plan the best places to take samples. We would choose a given number of sample

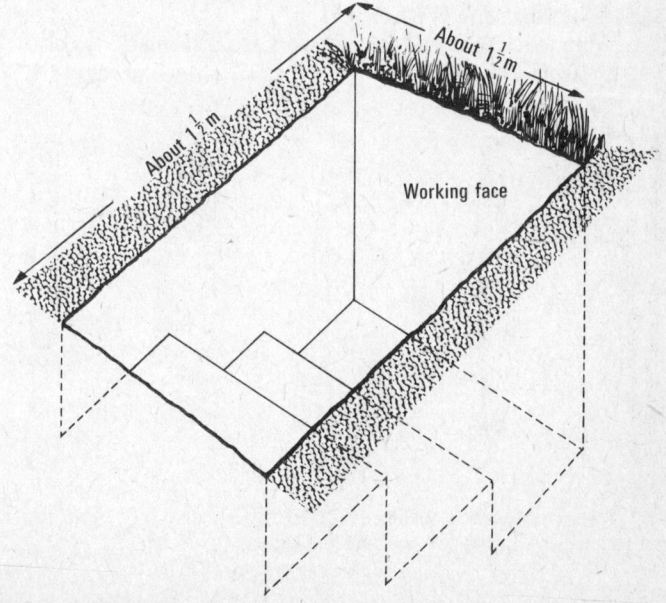

Figure 1.1.   Soil survey pits allow the detailed examination of a soil profile. Steps are necessary only for deep pits.

points in each of the different units shown on the map. However, it is usually the other way round—we have to study soil samples which will enable us to draw the map. If there are topographic maps or aerial photographs, sites will be chosen where they seem to be near the middle of a uniform piece of ground. We will avoid sites near sharp changes in topography or where they are unlikely to be typical. Without any maps or photos on which to plan sampling, a rectangular grid is used.

Especially when doing reconnaissance surveys it may be time-saving to sample where access is easy along a road, railway, or river, but this must be supported by grid sampling because these natural lines of communication may not be typical of the whole area. In dense vegetation it may be necessary to clear 'traces' or sight lines for the grid sampling.

The best way for the surveyor to study the soil at each sampling site is in an excavated pit. It should be big enough so that the surveyor can work in comfort, and extend down to about 1·5 m or to any lesser depth where there is a layer which would impede plant growth. The working face of the pit should be nearly vertical, and the excavated soil kept away from that end (Figure 1.1).

Digging pits takes time and is hard work, so information is also obtained by using soil *augers*. Several types are shown in Figure 1.2.

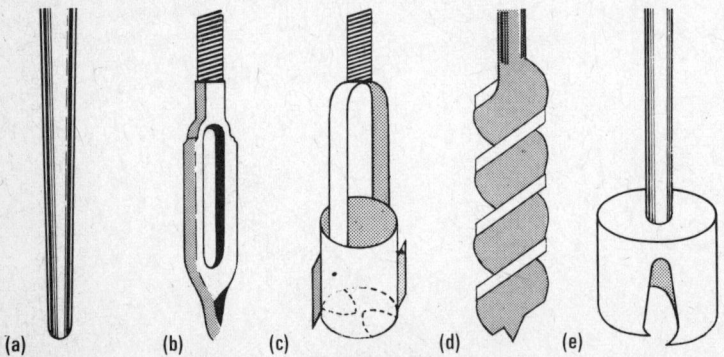

Figure 1.2.   Augers for soil examination: (a) probe; (b) Dutch; (c) Jarrett; (d) screw-flight; (e) post-hole digger.

The probe is quick and easy to use in soft soils but cannot be used in very hard or stony soils. The Dutch type gives a bigger sample but disturbs it, as does the Jarrett. The screw-flight auger is light to carry and easy to use, but gives a very small sample badly disturbed. The post-hole digger is very hard work and not suitable.

1.1.2. Soil factors to be recorded

Having gone to the expense of getting a trained soil surveyor to an inspection pit it will make very little difference in time or cost whether he measures and records 10 items of information or 12. So the usual practice, very sensibly, is to record all the information, both that which is required now and also anything which might be useful at some future date. The data collection takes place mainly in the field but a few determinations have to be made in the laboratory. The recorded factors are briefly described here; books giving more detail are listed in the list of *Further Reading* (p. 219).

Soil profile. Examination of the vertical side of a pit shows how the soil is made up of layers of soil with different properties. This is called the *soil profile*, and the layers are called *horizons*, which are identified by a letter and a number. The main kinds of horizons are listed below:

O   horizons consist of organic matter on top of the soil;
A   horizons are the topsoil;
B   horizons are the subsoil;
C   horizons are the soil being formed from parent material;
R   is the bedrock.

The A and B horizons may be subdivided, so the range of possible horizons is O_1, O_2, A_1, A_2, A_3, B_1, B_2, B_3, C, R, some of these having further subdivisions. However, most soils have only a few of the possible horizons. The presence or absence and the thickness of the horizons provides the knowledgeable surveyor with a pictorial record of how the soil was formed. The profile is therefore important in pedological surveys, but less important in surveys for agricultural development which put more emphasis on the agricultural properties.

Depth of soil and limiting horizon. In agricultural surveys the important thing about the profile is how much soil is available for plant roots to grow in, and this is called the effective depth. The lower limit to root growth is the boundary which defines the effective depth and may be laterite, parent rock, or a permanent water table. This is called the limiting horizon and its nature is recorded together with the effective depth.

Texture. Equally important for agricultural purposes is the texture, that is, the proportions of sand, silt, and clay. Each of the textural classes such as sandy clay, silt loam, etc. is defined in terms of the percentage of each size of particle. There are several such definitions and a simple and useful version is shown in Figure 1.3.

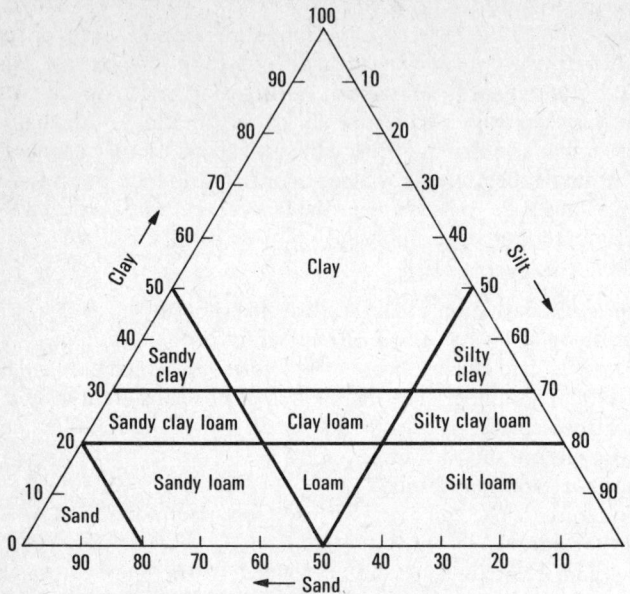

Figure 1.3.   Classification of soil texture.

Moisture-holding capacity. This is the amount of moisture in the soil which is available to plants. It is expressed as a proportion, such as millimetres of water per metre depth of soil. Not all moisture in the soil is available to plants. When the soil is saturated some will drain away before the plant can use it, and when the soil is drier than a certain point (called the *wilting point*) the plants cannot draw out the last remaining moisture. The moisture-holding capacity is very important for irrigated soils, and is discussed more fully in Chapter 6.
Structure and consistence. Soil structure is the way the individual soil particles are joined together to form larger aggregates. The shape and size, and the amount of aggregation can be assessed.

The strength of the cohesion between particles is called the *consistence*. Consistence may be measured on dry, moist, or wet soils, and the cohesion is quite different in each of these cases.

While the concept of these aspects of soil structure is clear, they are difficult to define precisely and accurately, and even more difficult to measure quantitatively.

Colour. The colour of soil is easy to observe, and a standard set of colours specially designed for soils have been accepted into international use. These are Munsell Colour Charts and are a systematic arrangement of all colours likely to occur in soils. The arrangement is based on three components; *hue* which is the dominant spectral colour, that is, position in the spectrum, *value* which is the brightness of the colour, and *chroma* which is the purity of the colour, that is, amount of included greyness. To use the charts, a moistened soil fragment from the upper subsoil is matched with the numbered colours through circular holes in the colours on the chart.

Measuring the soil colour by this means is straightforward, but interpreting the significance of colour is more complicated because a colour can indicate different things in different situations. In very general terms, reddish soils are usually well drained and well aerated, since the colour comes from unhydrated iron oxide. Yellowish soils resulting from hydrated iron oxide show that the soil is saturated for considerable periods. Black and dark grey soils are usually badly drained, and usually clays. Bluish tones in a dark grey soil show that the soil is waterlogged most of the time.

Permeability. Permeability is the rate at which water can move through the soil profile. Since this will vary at the different horizons, we have to specify the depth referred to, and the upper subsoil is usually chosen. The seven-class system of the U.S.D.A. is very widely used, sometimes with only the odd-numbered classes, 1, 3, 5, 7, being used for field assessments. The observation of the rate at which water is absorbed into a small lump of soil (called a ped) is sufficiently precise for allocation into one of these four groups, but not to differentiate between all seven classes. This requires more accurate measurement in the laboratory. The flow rates for each class are shown in Table 1.2.

Closely related, but not quite the same thing, is the *drainage*. The difference is that permeability is a measure of how water *could* pass through the soil, while drainage is a measure of how it *does* move. A soil of high permeability could be very badly drained if it is on top of an impermeable substratum. An assessment which is useful in conditions of seasonal rainfall is the length of time the land is waterlogged.

Table 1.2
Classes of soil permeability

Class	Description	Rate of flow†
1	Very slow	Less than 1·25
2	Slow	1·25 - 5
3	Moderately slow	5 - 20
4	Moderate	20 - 65
5	Moderately rapid	65 - 125
6	Rapid	125 - 250
7	Very rapid	Over 250

From U.S.D.A. Soil Survey Manual, Agricultural Handbook 18.

† The rate of flow in millimetres per hour through saturated undisturbed cores under a head of 12·5 mm of water.

All the previous soil features are *morphological*, that is concerned with the form of the soil. There are some other features which will be relevant when we come to decide how to use the soil.

Slope. The slope of the land is measured and expressed as a percentage, that is, the vertical rise per 100 horizontal units. For gentle slopes suitable for arable farming the 100 units may be measured down the slope instead of horizontal without the error being significant.

Erosion. A subjective assessment is made of the amount of erosion which can be observed to have taken place previously, whether man-made or natural. The four-class scale used in shown in Table 1.3.

Parent material. It may be helpful to know the kind of rock from which the soil has been derived. Although this information is not directly usable, it may provide clues to possible chemical or physical soil conditions.

Chemical factors. Agricultural uses of the soil will be heavily affected by its chemistry, but most of the chemical factors require laboratory analysis. The most useful items of information are the amounts of the major plant nutrients (nitrogen, phosphorous, and potassium), the organic matter content, and the salinity or alkalinity. For all these, samples would be sent to the laboratory. Only one chemical property is easily measured in the field and this is the degree of acidity or alkalinity, called the *soil reaction*, or pH.

Optional soil features. Some soil properties are only recorded when they are so pronounced that they could affect the choice of land use. Examples are poor drainage, indicated by W_1, W_2, or W_3, and the tendency to form a surface crust which could hinder the emergence of seedlings, shown as t_1, t_2, or t_3.

Table 1.3
Classes of erosion

Class	Description
1	No apparent, or slight, erosion
2	Moderate erosion: moderate loss of topsoil generally and/or some dissection by run-off channels or gullies
3	Severe erosion, severe loss of topsoil generally and/or marked dissection by run-off channels or gullies
4	Very severe erosion: complete truncation of the soil profile and exposure of the subsoil (B horizon) and/or deep and intricate dissection by run-off channels or gullies

From U.S.D.A. Soil Survey Manual, Agricultural Handbook 18.

1.1.3. Soil codes for mapping

The soil surveyor in the field studies the soil pit or the auger hole and makes an assessment of most, if not all, of the factors discussed in Section 1.1.2. He will almost certainly do this systematically by completing a printed form or using a specially prepared record book. A very large amount of data will be collected, and the question is how this should be presented to the man who is going to use it. The usual way is on soil maps which summarize the essential information, while the fuller detail can be referred to in a written report. The form of the map will depend upon the use. A map designed to emphasize the pedological features of the soils would concentrate on showing the soil series, that is, groups of soils with similar profiles, whereas the map to help plan arable farming would concentrate on different features.

Some kind of shorthand code is required to record simply and briefly the information which the soil surveyor has measured in the field, and this is done by means of a soil code which can be written directly onto a map or aerial photograph. Many different codes are used in different classification systems, and some examples are given in

(a) *Code used in Rhodesia*

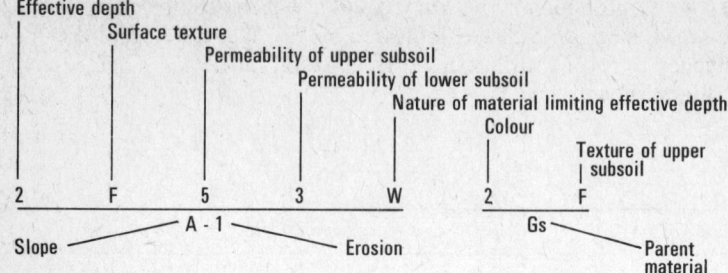

(b) *Standard code used in the Phillippines*

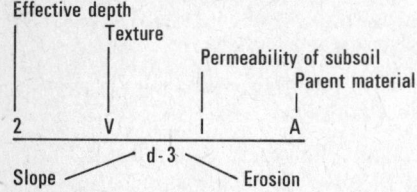

(c) *Phillippines Soil Type Code*

$$\frac{95}{d\text{-}3}$$

where 95 denotes a particular soil type, that is a textural classification within a soil series.

(d) *U.S. Department of Agriculture Series Code*

$$\frac{3401}{B\text{-}2}$$

where 3401 denotes a particular series, in this case Kalamazoo sandy loam.

Figure 1.4.   Some examples of soil mapping codes.

Figure 1.4. In order to record as much information as possible in the least space, each of the factors which have been measured is recorded as a number or letter which represents a single point on a scale. Thus the permeability is shown as the number of the class of permeability in the seven-point scale of classes shown in Table 1.2. Erosion is shown as one of the four classes of erosion in Table 1.3. For some factors the code consists of letters because these can be chosen so that they are meaningful and so easier to remember. The nature of the limiting horizon is an example of this, shown in Table 1.4. Another factor which lends itself to a letter code is the parent rock material. A small number of two-letter symbols (like GR for granite, GA for gabro) will usually describe all the most commonly occurring types of parent rock.

Table 1.4
Code symbols for limiting horizon

Symbols		Description
C		Relatively impermeable clay which may or may not show signs of regular waterlogging
H		A subsoil horizon that is sufficiently hard, dense, and compacted to prevent normal root development
	H1	Permeability not severely restricted
	H2	Permeability severely restricted, or relatively impermeable (generally deflocculated)
L		Laterite, wholly or partially indurated
	L1	Fractured, and permeability not severely restricted
	L2	Not fractured, and permeability severely restricted or relatively impermeable
M		A horizon (soil, gravel, or weathering rock) that shows signs of regular waterlogging which is likely to prevent normal root development for significant periods during the growing season
	M1	Composed of material that is not itself of severely restricted permeability
	M2	Composed of material that is of severely restricted permeability or relatively impermeable
Z		Gravel
	Z1	Loose and permeable with little or no soil; or moderately tight, but permeability not severely restricted
	Z2	Very tight and compacted, somewhat cemented and permeability severely restricted
W		Weathered, or partially weathered rock; or relatively unweathered rock that is well fractured; permeability not severely restricted
R		Rock, hard or weathered and not well fractured; permeability severely restricted or relatively impermeable

From Dept. of Conservation and Extension, Government of Rhodesia, Conservation Officers Handbook.

Table 1.5
Code symbols for effective depth

Symbol	Description	Range of depth (mm)
1	Deep	More than 1500
2	Moderately deep	1000 - 1500
3	Moderately shallow	500 - 1000
4	Shallow	250 - 500
5	Very shallow	Less than 250

From U.S.D.A. Soil Survey Manual, Agricultural Handbook 18.

Table 1.6
Code symbols for classes of land slope

Symbol	Range of slopes (per cent)
A	0 - 2
B	3 - 5
C	6 - 8
D	9 - 12

From U.S.D.A. Soil Survey Manual, Agricultural Handbook 18.

Table 1.7
Code symbols and definitions of soil texture

Symbol	Texture	Definition
A	Sand	More than 85 per cent sand
X	Loamy sand	80 - 85 per cent sand
B	Sandy loam	Less than 20 per cent clay; 50 - 80 per cent sand
C	Sandy clay loam	20 - 30 per cent clay; 50 - 80 per cent sand
D	Clay loam	20 - 30 per cent clay; less than 50 per cent sand
E	Sandy clay	More than 30 per cent clay; 50 - 70 per cent sand
F	Clay	30 - 50 per cent clay; less than 50 per cent sand
G	Heavy clay	More than 50 per cent clay

From Dept. of Conservation and Extension, Government of Rhodesia, Conservation Officers Handbook.

Other examples of code symbols are effective depth (Table 1.5) and slope (Table 1.6) where each number represents a defined range of numerical values. The code for texture (Table 1.7) uses letters, but they are alphabetical and do not relate directly to the names of the textures.

1.1.4. Soil survey from aerial photographs

The use of aerial photographs, often viewed in pairs under a stereo-scope to give a three-dimensional picture, can greatly help soil surveys but it is important to recognize what can be seen and what cannot. It is not possible to directly recognize and identify soil types on aerial photographs. We can literally see and recognize a mountain or river, but we cannot see the soil particles to determine the kind of soil. Instead we recognize a particular pattern on the photograph, and we also consider the slope, the vegetation, drainage, and a host of other clues. Together these suggest that the soil will be a certain type which is familiar because we have been and studied it on the ground. The ground work, correlating what we see on the photographs with what we can measure in the field, is essential. Sometimes variations in the soil are masked by vegetation, cultivation, or urban development. But even here the photographs help to decide where to dig inspection pits and to serve as a map on which to record soil data.

The main features of photographs which give information are tone, texture, and contrast.

Tone. The tone of a photograph is the shade of greyness, and depends upon the amount of light reflected by the ground surface. This is nothing to do with the colour as it would be seen on the ground. A dry, white, sandy soil will appear on the photograph as an almost white tone, but so does a highway although it is really almost black when seen on the ground. Soil moisture affects the tone on photographs with dry soils appearing lighter than moist soils.

Texture. This is the over-all appearance on the photographs of an area which is made up of a large number of separate features too small to be seen individually. We cannot recognize individual trees on small-scale photographs, but the texture of woodland or forest can be recognized. We have to know from ground studies what the different textures are representing. Texture and tone are closely related and the interpreter considers them together.

Contrast. Contrast is the difference between the different shades of greyness. A high-contrast photograph has a big difference between the very dark and the very light. A low-contrast photograph has mainly middle shades of grey. Contrast on the film negative is affected by the light when the photograph was taken, which depends upon time of day and time of year, and by the kind of film. The contrast of the photographs can be controlled during printing to give high or low contrast as required.

The sequence of interpreting photographs. First we must know, from ground studies, what soil types occur in the region which is being studied. Perhaps 3, 4, or 5 kinds can be identified. Next, by studying photographs of the areas where the soils have been identified, we learn what each of these soils looks like on the photograph and how to tell it apart from the others. After this we can look at photographs of other areas and see where an appearance which we recognize indicates one of our soil types. The number of soil types used for this exercise is limited by the number of different patterns which can be identified on the photographs, and this is usually up to about five. Further classification into subdivisions would have to be done on the ground.

1.2. Land classification

A survey consists of collecting facts together. The purpose may be to collect facts for a particular purpose, but it might also be to collect useful information without knowing exactly how it will be used later. Soil surveys can be of either kind.

Soil classification and land classification are different. Both of them make sense only when they are done with some particular objective. The process is to take the collected facts and put them into some sort of logical pattern, a kind of filing system so they can be located, summarized, tabulated, or analysed. But we need to know what use we are going to make of the facts before we can design a filing system which will give us the facts in usable form. The purpose of the classification will also tell us which facts we need to include, for there is no point in including irrelevant facts.

The test of a good classification system is that it must be able to start with the measured facts, and from this tell us which class to put it in. This requires clear rules, and any person using the system should get the same result. Too often a classification consists of a list of the properties found in each class, without explaining what to do with a set of data which does not fit exactly into any of the categories. Personal interpretation may be admissible for surveys when all the field surveyors are highly trained and experienced, and would all arrive at the same answer. But in countries looking for rapid agricultural development what is needed is to cut out all personal or subjective judgements, and provide a set of clear rules that anybody can understand and follow and get the same answer.

1.2.1. Special-purpose classifications

The simplest land classifications are those which record the existing situation—for example, present land-use. In this case we only need to decide how to collect the data and what classifications to use. The data could be collected by teams of field workers, inspection of aerial photographs, or perhaps in a developed country by a postal question-naire or from study of tax records. It is necessary to decide how many different classes will be used and what they will be, for example, will all arable land be shown as a single class? or should it be subdivided into different kinds of cropping? We have established the first principle—that a classification system must have a purpose—and now come to the second—that the availability of the data, and the filing system we are going to put it into, both have to be studied together at the beginning of the project. We cannot start collecting the data until we know whether we need areas of arable or areas of each crop, but we cannot decide the classification system until we have at least some idea of what the data is going to be.

The solution to this problem is to take alternate steps along each path. As an example, for a classification of physical topography using aerial photographs, the main stages would probably be as follows.

1. In the office a first interpretation to identify the main classes seen in the photos and to get an idea of how much there is of each class.
2. A field reconnaissance to see on the ground what is represented by the different forms on the photograph. A first draft of the classification system is set up as a map legend, that is, a description of each class which will be shown on the map.
3. Sample areas are then classified using the appropriate technique. This may be entirely air-photograph interpretation, or field study, or most likely a combination of both. The results of the sample survey will almost certainly bring to light new infor-mation, for example, what was provisionally classified as a single unit turns out to have two different kinds which should be recorded separately. Or one category turns out to occur in such small amounts that it is not worth recording it. As a result of the sample survey a new improved legend will be drawn up.
4. The main survey then takes place, recording in the field and on

photographs the location and boundaries of each of the classes in the legend.

5. Finally the result will be produced as a map, probably with supporting data tables.

Note that the purpose of the classification affects every stage. If the topographical classification is going to be used to plan agricultural development, the reconnaissance will concentrate on classifying possible farm land, and all the steep mountainous country may be pushed on one side in a single classification. If, however, the purpose is mineral exploration, then there will be geological studies later, so it would be important to study whether the steep mountainous country is all the same geologically or should be divided into subclasses. At all the other stages the objective of the classification will influence the decisions, right up to the question of what scale the final map should be.

It cannot be too strongly emphasized that there is no such thing as an all-purpose land-classification system. Land classification only makes sense when it is directed towards a particular purpose, and the more accurately that purpose is defined the more useful will the classification be.

In some cases the purpose can be clearly defined, but to achieve it requires more complicated inputs than the simple recording of observed facts. Let us look at an example that is of great importance in most developing countries. We wish to classify the land according to its suitability for irrigation development. The measurable facts are easy— the soil type, depth, moisture-holding capacity, and so on, and from there it is easy to classify according to the *suitability of the soil for irrigation*. We can also work out the areas where water is available, or could be made available, and this can be superimposed on the soils data to give a measure of the *practicability of irrigation*. But what we really want to know is whether it is going to be economically sensible to develop irrigation—that is to say, the real objective is a classification of the *profitability of irrigation*. This is much more difficult because we now have to include things like the cost of pumping the water and the probable sale value of the crop, and these are not simple measurable facts.

There are two approaches to this problem. One is to set up a classification for a particular project where the local costs, sale prices, and other similar factors can be estimated and included in the classification of profitability. This will work only for that particular time and place and must not be extrapolated. For instance, if cheap electric power is brought into the area this could lower pumping costs and change the relative difference between high land and low land. Another example is land which is fine for profitable irrigation of fresh vegetables if it is close to a city, but quite unsuitable in a remote district with no market outlets.

The other approach is to use the fixed measurable facts in a classification based on what is physically possible, and then superimpose a second set of criteria based on the variable factors. In other words, we first classify according to what is possible because of the permanent facts, then we modify the classification according to what is practicable considering the present circumstances.

When we come to classification according to possible land-use or possible agricultural development, the picture is even more complicated. Now even the basic facts are not always permanent. There is a piece of land which is so badly drained that at the moment it is useless for farming. But if our objective is to classify potential land-use we should also consider what it would be like if we drained it. Other land might be so steep that erosion would preclude using it, but we should consider whether it could be bench-terraced. If land is low in fertility, that too could be changed. So classifying possible future land-use has to include both the present conditions and possible changes.

1.2.2. Capability classification

In most developing countries the purpose of land classification is influenced by two factors. First, the desired objective is to achieve the maximum possible agricultural production, and this means using the land as intensively as possible. Second, most developing countries are in geographical and climatic situations where soil erosion is a major problem, and development has to be achieved without increasing the risk of erosion. Classifications have been developed for this situation and are called *Land Capability Classification*. The name is not the most appropriate because 'capability' is often confused with 'suitability', and the name does not explain the very clear objective, which is to classify land according to its capability for intensive agriculture without increasing the risk of erosion.

There is no single universal Land Capability Classification system, for in every country or region there are different factors which should be allowed for. The soils and climate will vary and so will social customs, land tenure, economics—and all of these may affect the choice of the best land use. However, for dryland farming on moderate slopes, the eight-class system originally developed by the United States Department

Table 1.8 Criteria for land capability classes

Land Capability Class	I	II	III		IV	
Permissible slope	0 - 2 per cent	2 - 5 per cent	0 - 5 per cent	5 - 8 per cent	0 - 8 per cent	8 - 12 per cent
Minimum effective depth (Texture here refers to average textures)	1 m of CL or heavier	500 mm of SaCl or heavier	500 mm of S or LS 250 mm of Sal or heavier	a. 500 mm of SaCL b. 250 mm of CL or heavier	250 mm of any texture	250 mm of SaCl or heavier
Texture of surface soil	CL or heavier	Sal or heavier S, or LS if upper subsoil is Sal or heavier	No direct limitations	a. Sal or heavier b. CL or heavier	No direct limitations	Sal or heavier
Permeability 5 or 4 to at least . . .	1 m	500 mm	No direct limitations	No direct limitations	No direct limitations	
Nor worse than 3 to . . .	1 m	1 m	1 m or 500 mm if average texture is CL or heavier	1 m		
Physical characteristics of the surface soil. Permissible symbols	Not permitted	t1	t1 and t2	t1 and t2	t1 and t2	t1 and t2
Erosion – permissible symbols	1	1 and 2	1, 2, and 3	1, 2, and 3	1, 2, and 3	1, 2, and 3
Wetness criteria – permissible symbols	Not permitted	W1	W1	W1	W1 and W2	W1 and W2

S = sand; Sal = sandy loam; LS = loamy sand; CL = clay loam; SaCl = sandy clay loam.

From Dept. of Conservation and Extension, Government of Rhodesia, Conservation Officers Handbook

of Agriculture Soil Conservation Service will serve in most situations with some modification for local conditions. A different system is required for the humid tropics where the emphasis on terracing and greater population pressure mean that steeper slopes are also brought into production.

A capability classification for moderate slopes. The system originally developed by the U.S.D.A. SCS has been modified for use in tropical and sub-tropical Africa, and used successfully in many countries. It consists of collecting the data described in Section 1.1.2 and coding it as in Section 1.1.3. From this information it follows that any piece of land falls into one of eight classes, four arable and four non-arable. The criteria for the four classes of arable land are more detailed than for the non-arable classes, where slope alone usually determines the class. The details of the arable classes are shown in Table 1.8. However, the practical application of capability classification means working from the soil data to the class, not the other way round. We are looking for the most intensive use, that is, the class with the lowest number, so the method is to test the measured data against the requirement of each class starting with Class I. If the data fails to meet any part of the specification for Class I we try it for Class II. If it will not do for Class II we try it for Class III, and so on. In practice this can best be done by using the systematic procedure shown in Figure 1.5. The main factors are considered in the right order moving down through the chart, starting with slope and finishing at a class. The minor factors shown on the left of the diagram are then checked to see if they will down-grade the classification. An appropriate level of soil-conservation measures is recommended for each class, and these are discussed in Chapter 8.

A capability classification for the humid tropics. When there is plenty of gently sloping land available for agriculture it is practical to impose fairly tight restrictions on the maximum slopes which are considered suitable for cultivation. This is the situation in the United States, where the land capability classification was developed, and in many other countries where it has been successfully applied—in Africa, Australia, and South America. But in parts of the humid tropics there is another combination of circumstances—high population pressures so that every hectare has to produce food for a hungry person, no more undeveloped land, and plenty of labour. In this case arable farming can and should be extended to much steeper slopes by the use of various forms of terracing. The details of the required conservation works are discussed

Additional requirements

*Factors affecting
cultivation*

g
b   Downgrade Class I to II
o
s
v   Class VI
r

Permeability
    3 to 20 — otherwise class IV
    Not applicable to basalts
    or norites

Erosion
    Class   I : 1
            II : 1 ; 2
            III : 1 ; 2 ; 3

t factors
    Class II : t1
          III : t1 ; t2
          IV : t1 ; t2

Wetness
    Class II ; w1
          III : w1
          IV : w2
          V : w3
w2 downgrades Class II and III
to IVw unless the land is
already Class IV on code, in
which case it remains as Class IV

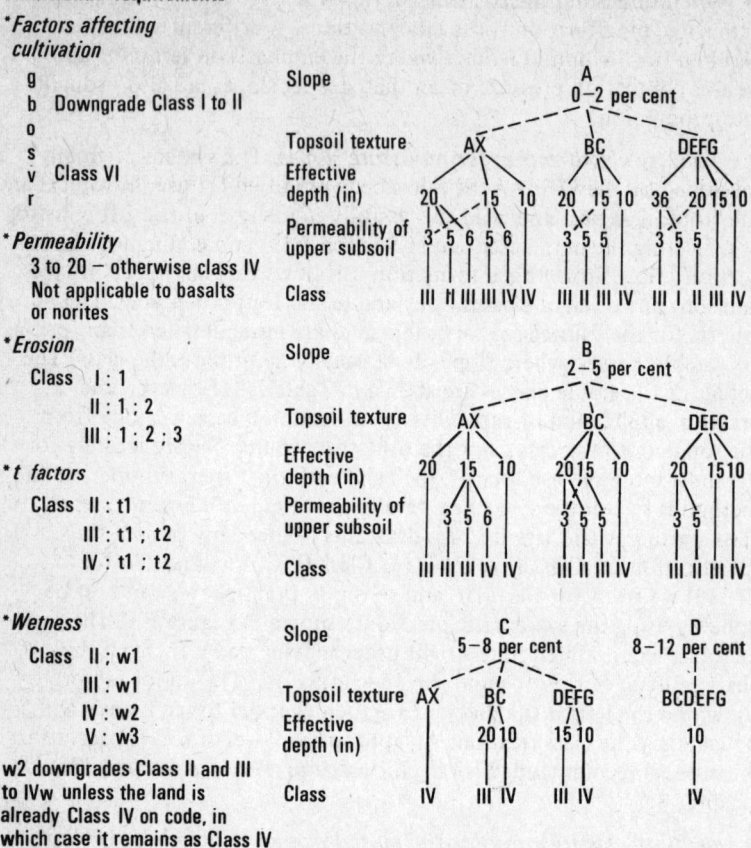

*Any land not meeting the minimum
requirements shown on this sheet is Class VI

Figure 1.5.   Systematic determination of Capability Classification (redrawn from a chart of the Planning Branch, Dept. of Conservation, Federal Government of Rhodesia and Nyasaland).

in Chapter 8. A capability classification scheme for this situation is shown in Table 1.9. The principle is the same, that is, to recommend the most intensive form of production together with the conservation

Table 1.9
Criteria for land capability classes in the humid tropics

Land class	Maximum slope degrees (Note 1)	Minimum soil depth (mm) (Note 2)	Conservation Treatment (Note 3)	Maximum intensity of land-use
1	7		0-2° Contour cultivation	Any
			2-7° Channel terraces	Any
2	15	1000	Bench terraces	Close-cover crops and semi-perennials
3	20	500	Step terraces or hillside ditches	Tree crops with ground cover
4	25	500	Step terraces or hillside ditches	Tree crops with ground cover
5	33	250	Orchard terraces or platforms	Tree crops with ground cover (no cultivation)
6	More than 33	-	None	Forest only

Notes 1. Equivalent slopes are 12, 27, 36, 42, 65 per cent.
2. Minimum soil depths are required when terraces are to be cut into the hillside.
3. The conservation treatments are explained in Chapter 8.

methods necessary to avoid undue erosion. Another difference is that effective use can be made of non-food crops and tree crops on land too steep for arable farming of food crops.

1.2.3. Classification for irrigation

Classification of land according to its possible use for irrigation has to take account of local factors, and so there is no system suitable for universal use. However, there is a system suitable for use as a base on which to superimpose the effect of local conditions. This system was developed in the United States by the Bureau of Reclamation, in connexion with their irrigation development schemes in the Western

United States. The basis of the classification is to assess the capacity of the land to earn money from irrigated farming, and this is used to determine the size of each holding and the charges for development and water supply. Thus one farmer might be given a slightly larger area to make up for it including some poor land, or another farmer might be on high land and so charged less for the supply of his water because he has to pay more for pumping it.

When irrigation development schemes require capital outlay which has to be recovered from the users this provides a way of sharing out the costs among the users. It can also be used to share out the running costs of a scheme put in for existing land owners. There are six classes in the system:

Class 1 has the highest level of irrigation suitability, hence the highest payment capacity;

Class 2 has intermediate suitability and payment capacity;

Class 3 has the lowest suitability and payment capacity;

Class 4 designates special-use classes such as fruit 4F, and also land which could be irrigable after some special problem or deficiency has been corrected—for example, a salinity condition which can be overcome;

Class 5 is used as a temporary designation for lands requiring special studies before a final choice of class is made;

Class 6 is land not suitable for irrigation development.

As in the land capability classification, the basic physical factors are first used to suggest a probable class, then secondary factors are checked to see if they down-grade the class. The physical factors are similar to those used in capability classification, and they are recorded in a coded layout which is also similar. An example of the data

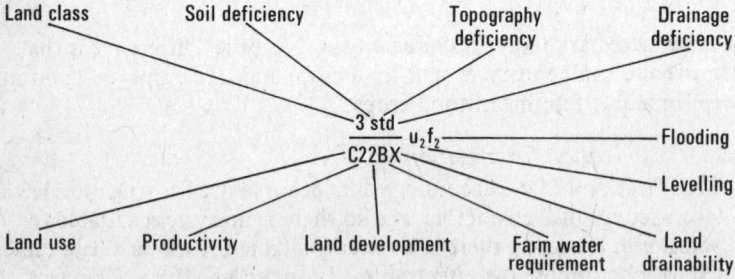

Figure 1.6.   The soil code for irrigation surveys used by the United States Bureau of Reclamation.

Table 1.10
Code symbols for additional data used for irrigation soil survey

Salinity, alkalinity, problems, and reaction	Drainage and flooding, problem extent, subsurface and surface
Symbol†	Symbol
S_1 Moderately saline soil EC‡ 4 - 8, pH below 8·5	Dl Adequate drainage depth and capacity available to solve a drainage problem locally
S_2 Severely saline soil EC above 8, pH below 8·5	Dp Drainage problem requires construction of outlet facilities for a considerable part of the project
AS Saline-alkali soil EC above 4, pH about 8·5	Fl Flood danger, but adequate outlet capacity available to solve the problem locally
AL Alkali soil, EC below 4, pH above 8·5	Fp Flood danger requiring large-scale measures affecting at least the whole project area
AC Acid soil, pH below 6·0	W Water table (give depth to water table as subnumber in feet or metres)

Topography	Land-levelling requirement for irrigation and drainage §
Symbol	
L Long, continuous slopes facilitating long irrigation runs	1 Light (50 - 150 m³/ha)
F Flat terrain fairly well graded	2 Moderate (150 - 300 m³/ha)
R Rolling terrain with short and alternating steep and flat slopes	3 Heavy (above 300 m³/ha)
FR Flat but rolling terrain requiring considerable land-levelling	Where less than 50 m³/ha are required no symbol need be given
B Broken terrain cut at frequent intervals by gullies, walls, etc.	
T Terraced terrain	

Slope percentage is written before symbol; example 2 - 3L² means 2 - 3 per cent continuous slope requiring 150 - 300 m³/ha of levelling.

From U.S.D.I. Bureau of Reclamation.

† Salinity symbol is written as prefix to the digit representing the affected horizon (for example, AL 2).
‡ EC = electric conductivity in millimhos per centimetre at 25°C.
§ Write land-levelling degree digit in exponent position to the letter representing topography factors.

Table 1.11
Physical criteria for irrigation classification

		CLASS 1	CLASS 2	CLASS 3
Texture	lightest heaviest	Sandy loam Silt loam	Loamy sand Silty clay loam	Loamy sand Silty clay
Minimum depth		lm fsl† or sl	600 mm fsl-cl or 750 mm ls	450 mm sl-cl or 600 mm ls
Size of irrigation field Length (m) Size (ha)		120 3	100 2	50 1
Earth-moving Minimum excavation for drains (m^3/ha)		400	800	1400
Maximum land-levelling (m^3/ha)		400	800	1400
Average cut and fill		75 mm	150 mm	250 mm

From U.S.D.I. Bureau of Reclamation

Notes
1. All classes must meet the specifications on permeability, salinity, and alkalinity.
2. Special local criteria may apply, for example, number of trees per hectare.
3. Any other factors which affect either the profitability of irrigated crops or the cost of development may be included.
4. Special problems are indicated by suffixes: s for soil, t for topography, and d for drainage.

† fsl = fine sandy loam, sl = sandy loam, cl = clay loam, ls = loamy sand.

symbols is shown in Table 1.10 and an example of an irrigation soil survey code in Figure 1.6.

The physical criteria for the three main irrigable classes are shown in Table 1.11. If subsidiary factors do change the class this is shown by adding a suffix after the class number—s for soil, t for topography, and d for drainage. Thus Class 2s means that it does not meet the Class 1 specification for soil, but does in other respects. A soil classified as 3std means it fails to meet the Class 2 requirement in all three categories.

2. Land development

2.1. Development planning

Whether land development is to consist of clearing new ground or the redevelopment in a new form of land already in use, the important thing is to have a comprehensive plan which includes all aspects of land-use. The details required will vary according to the size of the scheme. A development scheme for a whole region will include plans for things like transport and marketing, whereas these would not be appropriate in a plan for developing a few hundred hectares. However, the principle applies to development planning on any scale. It should be comprehensive even if it is mainly concerned with how-to-do-it techniques like the next topic of how to clear unused land and make it fit for agriculture.

2.1.1. Operation of machinery

Most land development schemes involve earth-moving and other heavy machinery, and a basic consideration is how the machinery should be owned and operated. The choice lies between the Government, the land-owner, contractors, or other schemes. This question will also be important when we consider other operations such as the construction of dams, conservation measures, drainage, and irrigation.

Government ownership. Several advantages arise from the use of fleets of heavy machinery owned and operated by the Government.

1. Frequently it is only Governments which have the capital (or the credit facilities) to finance the fleet, and the supporting spares and workshops.
2. Being independent of the need to make a profit gives a direct financial advantage, and also other benefits. For example, schemes which are uneconomic but nationally desirable can be undertaken, or urgent jobs can be given priority.
3. Government policies can be assisted by giving preferential rates of hire or priority treatment for any particular kind of farming which the Government wishes to encourage. Particular enterprises like irrigation, or fish farming, can be stimulated, if it is

Government policy to do so, by making the machinery more
readily available.

4. Similarly, a region can be assisted if a Government wishes to
   develop it, or it has suffered hardship like drought.
5. Another result of a flexible approach to costs is that Government
   machinery can undertake both profitable and non-profitable jobs,
   and balance the income from both. The private contractor costs
   each job separately so the land-owner whose land is less
   accessible, or whose site conditions are less favourable, has to pay
   a higher price. Government machinery can be charged at an
   average cost which evens out such differences and provides a
   service for everybody.
6. There are technical advantages in operating machinery in large
   groups. They may justify employing a resident engineer,
   specialized workshop facilities, or a better stores and spare-parts
   service, all of which should lead to more efficiency compared
   with smaller operations.
7. The Government can establish and maintain standards of work-
   manship, be a model employer, and provide training programmes
   to the general benefit of the country.

The disadvantage of Government-operated machinery is that the
administrative costs always seem to be so high that the final cost is
greater than when done by private enterprise.

Private contractors. The advantage of the private contractor is that
because he is in a competitive commercial business he has, or should
have, the most suitable equipment, and the knowledge and experience
of how to use it with maximum efficiency. The disadvantage is that
since the contractor is in business his profit margin has to be added to
the cost. His increased efficiency may, or may not, outweigh this extra
cost.

Land-owners or farmers. Few land-owners or farmers can afford to own
the larger items of specialized machinery, but as we shall see later many
development jobs can be done with farm-size equipment which can be
operated through the joint effort of a village, a district, or a co-operative.
The advantage of work being carried out by the occupier of the land is
that he will be personally involved and feel responsible. The job will
probably be much better done, and certainly better cared for afterwards, than if it were done by someone else.

The other advantage is that by carefully planning the timing the job
can be done cheaply by fitting it in with other farm work. Economists

use the expression *opportunity cost* to allow for the effect of what the equipment might be doing if it were otherwise employed. Because farm demands of labour and equipment fluctuate so much with the seasons, there are times when farm equipment and drivers are not needed for other jobs and so the opportunity cost is low.

2.2. Clearing and levelling

2.2.1. Clearing heavy forest

Clearing dense forest is expensive and laborious no matter what machinery and equipment is available. The over-all cost may be offset by the extraction of merchantable timber, and this should be included as part of the development plan. It can bring benefits in the form of roads and bridges built to get the timber out, which then remain for agricultural use.

Commercial timber exploitation needs to be controlled. Unscrupulous operators are likely to go through the forest selectively picking out the best trees and the highest-priced woods, and may leave a lot of second- and third-grade timber which is uneconomic to harvest on its own. A well-managed logging enterprise can make subsequent development easier and cheaper, but a bad one can leave the loggers wealthy but everybody else worse off. The problems which can arise from bad logging are leaving a tangle of carelessly felled other trees, excessive erosion from logging trails, and leaving badly sited roads and trails.

The amount of clearing will depend on the planned land-use. For intensive cropping, or irrigation land which needs levelling, complete removal of all the trees, stumps, and roots may be necessary. On the other hand, for some types of development such as oil-palm estates it is sufficient to fell all the trees but leave most of them on site to reduce the clearing costs. The lighter growth will probably be burnt, and the rest bulldozed clear of the harvesting paths and roads.

Clearing costs may also be reduced by leaving individual large trees if their removal would need special treatment. Leaving some shade trees may also be desirable if livestock are intended.

2.2.2. Machinery for clearing

Bulldozers are used more than any other machine for clearing trees by pushing them over. The overturning action tears out or loosens a lot of the roots, especially shallow roots. The overturning moment can be increased by raising the point of application of the push by lifting the dozer blade as high as possible, or by having a high raised peak on the blade, or by the use of hydraulic rams which apply the thrust high up the tree stem.

There are several special-purpose bulldozer blades (Figure 2.1). The 'stumper' has a small squat blade with teeth at the bottom for digging out tree stumps. The 'cutter' has a high pusher at the top and a straight blade at the bottom for cutting the roots. The 'Rome' blade gets rid of the tree but leaves the stump, and to do this it has a 'stinger' which splits the tree stem and a sharp horizontal blade which shears off the stem at ground level.

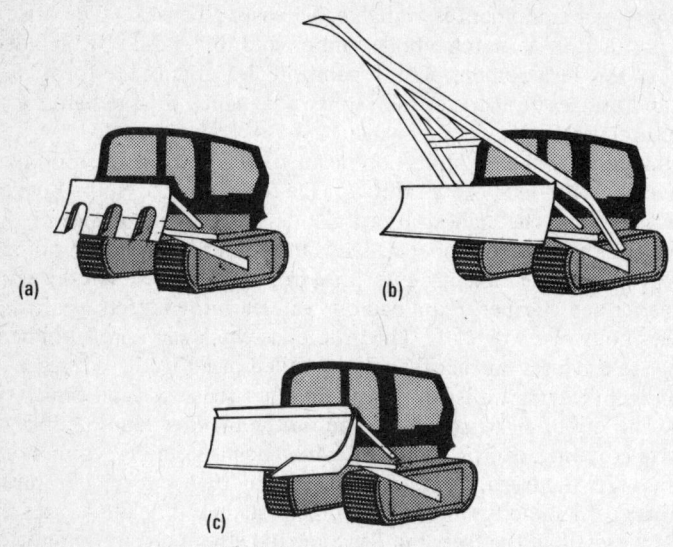

Figure 2.1.   Special bulldozer attachments for tree-clearing. (a) Stumper for grubbing out tree stumps. (b) Tree pusher and cutter. (c) Tree feller with stinger.

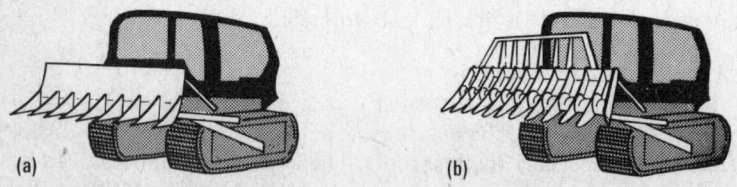

Figure 2.2.   Special bulldozer attachments for root clearing: (a) toothed bulldozers; (b) root rakes.

Figure 2.3.   A root plough for cutting tree roots and bringing them to the surface.

After felling and stumping the trees the next operation is to get out the roots. This is done with a toothed dozer blade, or a root rake, which is similar but has a hollow framework so that soil is spilled through the blade and only the roots retained (Figure 2.2).

If the roots have not been brought to the surface by the stumping and clearing, a root plough can be used. These may be front-mounted like a dozer, or rear-mounted like a tool-bar, and consist of a heavy flat blade drawn along below the surface to shear off roots and bring them to the surface (Figure 2.3).

The trees, stumps, and roots should be bulldozed into windrows for burning, after which the ploughing and discing can start. When the land being developed is prairie or treeless savannah it may be possible to use a 'prairie-buster' or 'digger' mouldboard plough. These are large, tough ploughs with good clearance to avoid choking with trash, and often have a 'stump-jump' device which is a spring-loaded release to prevent damage when the plough hits an obstacle. Where rocks and stumps are common a disc-plough will be more appropriate, and again heavy-duty land-breaking machines are available, such as the 'Rome' plough. The name plough is rather misleading for they have little resemblance to ordinary ploughs. They can weigh up to 4 or 5 tonnes, and have discs from 0·5 m to 1·5 m in diameter, and need a very large crawler tractor to pull them.

2.2.3. Clearing light woodland and bush

The cost of clearing is much less if the growth is light enough to be taken out in a swathe, instead of having to attack individual trees. The two basic approaches are either to pull out everything in one pass, or to mangle it up and then sweep it into piles for burning. It may be possible to start by burning off the standing vegetation, but unless this will give a good burn it is as well to clear first and then be sure of a good burn in piles.

The 'ball and chain' is a good one-pass operation. A very heavy steel chain (weighing up to 15 kg/m (10 lb/ft)) is attached at each end to large crawlers which move in parallel with the chain in a big loop behind and between them. To lift the chain up for better leverage, it is attached at midpoint to a large iron ball about a metre in diameter. A third crawler with a dozer blade is sometimes used as a follower behind the chain to give an extra push to bigger trees.

It may be easier to crush and break up the vegetation, and several machines are available and particularly suitable for less woody growth. The rolling brush chopper and the 'Holt' weed-breaker both give a chopping action as each cutting edge comes down.

A similar pulverizing effect is achieved by flails and rotary cutters. Flails have lengths of chain attached to a cylinder which rotates about a horizontal axis, and can be mounted ahead of or behind the tractor. Rotary cutters are rear-mounted and like enlarged grass mowers, with chains or swinging blades attached to a circular plate which revolves about a vertical axis.

2.2.4. Some general points about clearing

1. The sequence of operations required will vary according to the existing vegetation and the proposed land use, but some typical examples are given in Table 2.1.

Table 2.1
Clearing different kinds of vegetation

Heavy forest	Heavy bush	Light bush
1 Selective removal of timber	1 Ball-and-chain clearing	1 Burn off top growth
2 Clearing and stumping, rooting	2 Doze into windrows	2 Heavy-duty disc
3 Doze into windrows	3 Burn	3 Rake into windrows
4 Burn	4 Plough	4 Burn
5 Plough		

2. Clearing is dangerous for tractor drivers and proper precautions must be taken. Because of the danger of falling trees and whipping branches, tractors should have safety cabs and operators should wear protective clothing.

3. During the first clearing operation the operator cannot see ahead through the bush so tracked tractors are preferable to wheeled tractors, and front-mounted implements are preferable to towed implements. When wheeled tractors have to be used, steel wheels may be better than tyres.

4. The intended land-use will dictate how much clearing is necessary. Leaving some trees standing, or leaving the stumps and logs on site, can reduce the cost.

5. Follow-up treatment will be required to prevent re-growth from roots missed during the first clearing, and to prevent re-colonization.

2.2.5. Other methods for clearing

When machinery for land clearing is too expensive or not available, other methods must be considered. Hand labour can be made more effective by mechanizing part of the job, such as using petrol-engined chain saws, or clearing by hand but using tractors to haul off or stockpile for burning. Stumping of the largest trees can be done by hand, using pullers to gain mechanical advantage. The 'Trewella monkey winch' is a block-and-tackle puller used effectively in Africa. It is slow, and needs a lot of labour, but these are not always serious disadvantages.

Explosives can be useful for removing large stumps. The technique is to bore a hole with a soil auger so that the explosive can be placed right under the stump. Dynamite is the best explosive, and blasting is best done when the soil is wet. Blasting can also be used to split very large stumps. In this case a wood auger is used to bore a hole in the stump big enough to insert the dynamite sticks. Blasting should only be attempted by people properly trained in the handling of explosives.

Chemical poisons may be appropriate in some cases. Arsenical poisons are dangerous to man and beast, and so only suitable for spot application, but they are cheap and easily applied, for example, for pouring into tree stumps to prevent re-growth. Hormone growth-regulators of the 2-4-D type are used as blanket sprays, mainly on less woody crops. Because of the cost this is only likely to be practical for local applications, for example, the destruction of weed growth on road sides. Aerial spraying of mesquite (*acacia* species) is practised in the south west of America, where its encroachment into range land is a problem.

Fire can be a valuable tool in the management of savannah grassland and woodland. The tree and bush species which constantly try to encroach, to the detriment of the grass, usually have some degree of resistance to fire, and a hot, dense fire is needed to kill them. If uncontrolled fires are prevented, and the grass kept ungrazed for a time, there will be enough vegetation for the hot fire that is required.

2.2.6. Levelling for irrigation

Machinery. Heavy earth-moving machinery is discussed in Chapter 5 in connection with dam construction, but a brief mention should be made of the machinery used for preparing land for irrigation. All irrigation methods are made more efficient by some degree of land smoothing or levelling, and some require very accurate levelling in order to function at all.

Levelling for irrigation is different from most other earth-moving jobs in that the amount of soil to be moved is not large, but the depth of cut and the height of the fill have to be exact.

Bulldozers give the cheapest cost per cubic metre, especially for moving short distances, but lose efficiency as the distance increases. For hauls over 100 m it is better to pick up and carry using a scraper. The longer the haul the more likely it is that wheels will be preferable to tracks.

Scrapers give more accurate levels than dozers, but still not enough accuracy for irrigation so the final smoothing needs a grader or land plane. A self-propelled grader as used for road construction gives good results, especially when preparing long graded beds like border strips (Chapter 6). The most accurate levelling is achieved by a land plane which is like a grader with a very long wheel base, and which planes soil off the high spots and deposits it in the low spots. However, a land plane should only be used for the final smoothing after the main movement of earth has been done by dozer or scraper. The main features of a land plane are shown in Figure 2.4. It should be at least 8 m long, with the blade about two-thirds of the way towards the back, and the blade should be tilted forward at the top, not backwards like a dozer blade. The front is hitched to the hydraulic lift of the tractor and this gives control of the depth of cut, and allows the blade to be lifted clear for turning. The back wheels run within the planed surface.

Calculation of levels for irrigation. The problem is to work out where to cut high spots and where to fill low spots to give the required land slope with the least earth movement. The accuracy required depends on the irrigation method (Chapter 6), the greatest accuracy being required for border strips which need levelling to ±25 mm. In general, steeper grades need less accuracy than flat grades. More accuracy is required when the irrigation is to run on the contour than when it is down the steepest slope.

(a)

(b)

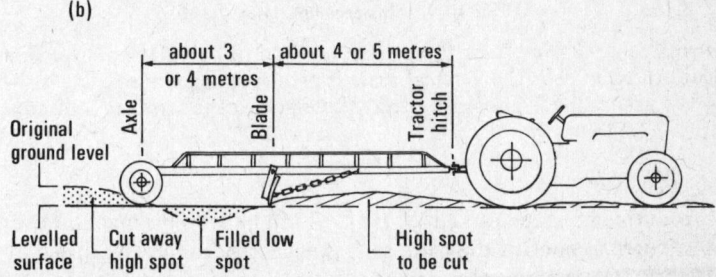

Figure 2.4.   A land plane is used for final accurate levelling of irrigation land.

Theoretically the volume of earth in the cut areas should equal the volume of the fill areas. However, in practice some soil is lost while being moved from cut to fill, and it is found from experience that the cut volume has to be greater than the fill volume by about 20 per cent.

Several methods are available for the calculation of cut and fill, ranging from simple trial-and-error methods to complete mathematical solutions by computer. The *profile method* is reliable and easy to use. It is particularly useful for levelling into graded strips which do not need to be all at the same elevation.

Straight lines are staked at intervals of 20 or 30 m, the lines running in the direction of the irrigation run. The ground elevation of each staked point is measured, and the profile along each line is plotted on squared paper (Figure 2.5). Possible grade lines are drawn until, by trial and error, the line which gives the required ratio of 1·2 cut to 1·0 fill is found. The depth of cut or fill at each point can be read off directly by comparing the chosen grade line with the plotted profile.

In the field the cut points are marked by tying a piece of red cloth to the stake; blue cloth is used for fill. However, the tractor driver cannot

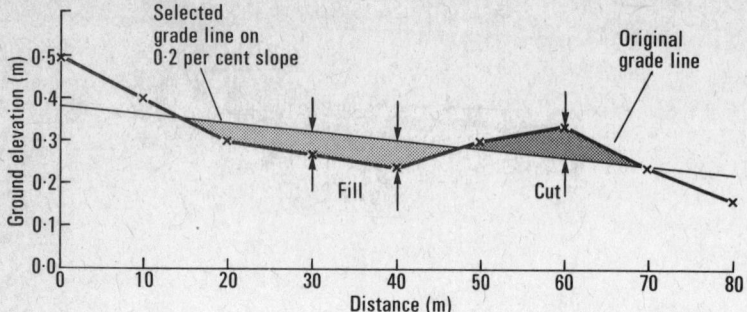

Figure 2.5.   Example of the profile method of calculating levels for irrigation. Note that the vertical scale is greatly exaggerated. The selected grade line is chosen by trial and error so that the ratio of cut to fill is 1·2 to 1·0.

tell from this how deep to cut or fill, nor can he stop his machine every few minutes to consult the plan, so it is necessary to have a ground assistant who reads the plan and signals the depths to the operator using clear signals of a pre-arranged code.

2.3. Land reclamation

2.3.1. Reclamation of waterlogged soils

To reclaim waterlogged soils they have to be drained, but the problem is to get men or machines on to the land to cut the necessary drains. Sometimes an improvement made outside the wet area will make sufficient difference to allow a direct attack on the problem area. For example, if the excess water comes from outside the area, then either surface or subsurface drains should be used to intercept the flow from outside. Another possibility is to increase the drainage of the wet area by clearing or lowering the bed of existing drainage channels, or by cutting new channels, or by pumping.

If this indirect approach is not possible, then the problem will best be tackled in successive stages. First the simplest possible system of open drains is cut, doing just enough in the bad conditions to get some water moving and start an improvement. When working conditions have improved, a second more detailed drainage system can be added, and after this has had its effect the real work of making the land productive can start.

For the first drains it may be necessary to dig them by hand, but there are special tools available. The 'water buffalo' is a crawler tractor with buoyant tracks nearly 2 m wide, which give support and traction in otherwise impassable swamps. It pulls a drainage plough like a large mouldboard plough which cuts an open drain and has an extension like a grader blade to push the excavated soil back from the drain. Some drainage ploughs are double-sided and cut a flat-bottomed ditch with the soil piled on both sides. Open drains cut half a metre deep and at intervals of up to 400 m will usually improve the land enough to be able to go in after one or two years with ordinary crawlers and excavate deeper drains at closer intervals of say 100 m between the first shallow drains.

If the tendency to poor drainage is likely to be a permanent feature, it will be necessary to put in a proper drainage system, and the various methods are discussed in Chapter 7.

2.3.2. Reclamation of saline soils

Salinity is having an excessive amount of dissolved salts in the water or the soil so that crop production is reduced or prevented. In mild cases the effect may be only a slight reduction of yield. In more serious cases the choice of crops may be limited to those which can stand some degree of salinity. In the worst cases nothing is able to grow. Table 2.2 shows a selection of crops which have different degrees of tolerance.

Salinity may occur naturally, or as a result of irrigating. Most water contains some dissolved salts, and when water evaporates from the ground only the water goes off, leaving the salts behind. In temperate climates these salts get washed out by water percolating through the soil, but in arid countries with high evaporation and low rainfall there is likely to be a progressive build up of salts. This is particularly marked in the case of enclosed drainage systems where rivers or streams continually bring in salts but there is no outlet except by evaporation. The Dead Sea is an example of this where the salinity of the water has built up to an extremely high level.

The same effect can also result from irrigation, for even if the dissolved salts are not enough to interfere with crop growth, they can build up in the soil over the years until they do become a problem. In some countries which have had irrigation schemes for hundreds of years, large areas of land have been abandoned because this has happened, for example, in India and Pakistan. This problem can be prevented or reduced by suitable management, although it is difficult to correct after it has happened.

Table 2.2
Crop tolerance of salinity

Low salt-tolerance	Medium salt-tolerance	High salt-tolerance
Field crops		
beans,	rye, wheat, oats,	barley
	sorghum, maize, rice	sugar beet
	sunflower	cotton
	castor beans	rape
Forage crops		
most kinds	sweet clover, Hubam	alkali sacaton
of clover	clover	
	most grasses	salt grass
	most grass crops when	*Cynoden* spp.
	taken for hay	*Chloris* spp.
		Agropyron spp.
		Barley (for hay)
Vegetables		
radish	tomato	asparagus
	lettuce	
	cabbage	
celery	cauliflower	spinach
	potatoes	
green beans	carrots	kale
	onions	
	peas	
Fruits		
all citrus fruit	fig	date palm
	olive	
deciduous fruits,	grape	
for example,	pomegranate	
apple pear, plum		
berries,		
for example,		
strawberries		

From U.S.D.A. Agricultural Handbook 60.

Different chemical salts cause different kinds of salinity, and require different treatments. The most common salts are the sulphates, bicarbonates, and chlorides of calcium, magnesium, and sodium. There are less common problems, although they can be just as serious, resulting from an excess of potassium, silicon, iron, or boron.

Sodium in particular has a special effect on the soil because it causes deflocculation, that is, a breakdown of structure. The two things which affect salinity are therefore the total amount of dissolved salts and the proportion of those salts which are sodium salts. Three main kinds of salinity occur:

saline soils have high total salts, but low sodium (sometimes called white alkali soils from the characteristic pale surface crust of salts); *sodic soils* have high sodium, but low total salts (sometimes called black alkali); *saline alkali* soils are high in both total salts and sodium.

This is shown diagramatically in Table 2.3.

Table 2.3
Soils with salt problems

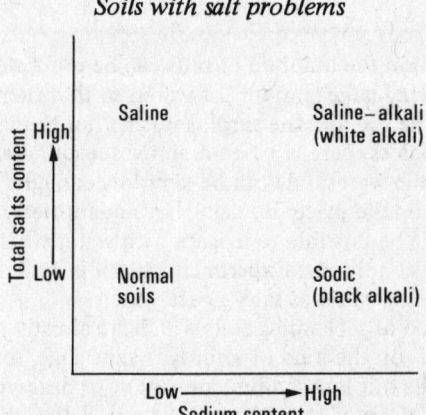

Reclamation treatments. As with so many agricultural problems, prevention is better than cure. The problem of salinity is less likely to occur if irrigation is only done with water of good quality containing small amounts of salts. However, chemical treatment of the water is impractical, so if poor water has to be used the right method of application is necessary.

Furrow irrigation should be avoided if either the soil or water is saline, because the salts are continually moved through the soil from the furrow to the ridge where they accumulate (Figure 2.6).

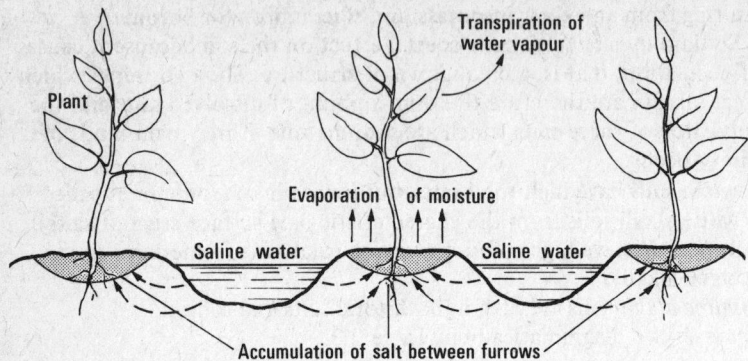

Figure 2.6.   Furrow irrigation is undesirable if the soil or water is saline.

In flood irrigation the build-up of salts can be prevented by periodically applying large amounts of water, so that there is more than can be stored in the soil and the surplus washes out the salts. This is called *leaching*. Unless there is a free-draining subsoil, drainage must be provided so that the watertable can be kept low enough for this leaching process to take place. By using large quantities of water on well-drained soils it can be possible to irrigate with water which would be unusable in other conditions. Experiments in India and in Israel have produced crops using water as salty as sea-water.

When it is a case of reclaiming soils which are already saline, the treatment depends on the kind of salinity. Saline soils, that is soils with excessive total salts but low sodium, are easiest to deal with as all that is required is lots of water to wash the salts out. Saline-alkali soils have high total salts and high sodium. Again leaching is the answer, but after leaching, the high sodium is likely to cause loss of structure, packing of the soil, and loss of infiltration. To prevent this the sodium has to be replaced by calcium. Additives applied to the soil to do this are described in the next paragraph. Sodic soils have high sodium but low total salts. Here replacing the sodium with calcium has to come first, so the technique is to do this with additives, and then leach out the sodium.

It is usually required to get rid of sodium and substitute calcium, and several calcium compounds can be used to do this. Calcium chloride is good but usually too expensive. Calcium carbonate is more readily and cheaply available as chalk, lime, or ground limestone, but it is only very slightly soluble so acts slowly. It is also unsuitable on soils which are already on the alkaline side. Calcium sulphate, called gypsum, is fairly common, cheap, and the most-used additive. It is applied in powdered form up to 12 tonne/ha (5 tons/acre), followed immediately by leaching. Sulphur is another cheap and popular additive. It is applied at up to 2·5 tonne/ha (1 ton/acre) and left for 3 months before leaching. The delay is needed for the sulphur to form sulphuric acid, then sulphates, which then act like gypsum.

2.3.3. Reclamation of eroded soils

Some land is destined by nature to be barren and subject to erosion and it would be impractical and uneconomic to convert it to productive farm land. But land which has slipped into this state through mis-use or neglect may be worth reclamation.

Eroded lands often have low infiltration, high run-off, high rates of erosion, and sparse vegetation, and it is not easy to identify which is the cause and which is the effect. If there is a reasonable amount of rainfall, a good starting point is water-conservation measures, so that there is more moisture for better growth and less run-off to cause erosion. Pasture furrows are small furrows dug on a level contour, at close spacing of from 5 m to 20 m (15 - 60 ft). Another way is to make shallow basins which also catch run-off and allow it to infiltrate. Range-pitting is the American name for doing this by machines.

Sometimes the infiltration can be improved by loosening the soil surface, especially when the soil is permeable underneath but has developed a hard surface crust. A shallow scarifying of the surface with a disc-harrow is often sufficient. Ripping or subsoiling should only be used when it is necessary to break through a hard pan or a non-permeable layer such as laterite.

When the basic limitation is lack of fertility, or a chemical deficiency, this has to be identified and corrected. Application of fertilizers or manure or chemical amendments may be the solution. Or perhaps the existing vegetation is unable to make full use of the moisture and plant nutrients which are already present, and a different species or a better variety of the same species could be introduced. In the United States some dramatic improvements of poor grazing land have been made by replacing the indigenous scrub vegetation with improved grass varieties.

3. Run-off and steamflow

Water is the key to agricultural development. The rate and the extent of development will depend on the available water resources more than on any other factor. It is therefore important to be able to measure or estimate water supplies. In order to plan water storage and irrigation, we need to know the *amounts* of surface run-off, and in order to design dams, drains, and bridges we need to know the *rates* of run-off.

3.1. Estimating rates of surface run-off

Many methods have been developed for estimating the probable maximum floods to be expected from small agricultural catchments. Two of these will be described which are simple and reliable.

3.1.1. The rational-formula method

The *rational formula* is

$$Q = \frac{C \times I \times A}{360},$$

where Q is the rate of run-off in cubic metres per second,

I is the intensity, that is the rate of rainfall, in millimetres per hour,

A is the catchment area in hectares, and

C is a dimensionless constant.

The formula was in fact derived originally in English units and owes part of its popularity to the fact that, when using the most convenient English units, C becomes dimensionless because of a fortunate numerical coincidence:

For rain falling at 1 in/hour on 1 acre,

43560 (acre to ft^2) $\times \frac{1}{12}$ (inch/hour to ft/hour) $\times \frac{1}{3600}$ (hours to seconds)
= 1·008 ft^3/s. which for all practical purposes can be taken as unity.

In English units the formula is

$$Q = C \times I \times A,$$

where Q is the rate of run-off in cubic feet per second,

I is the intensity in inches per hour,

A is the area in acres, and

C is the same dimensionless constant as in the metric formula.

Table 3.1
Gathering time for small catchments

Average gradient in catchment (per cent)	0·05	0·1	0·5	1·0	2·0	5·0
Maximum length of flow (m)	Gathering time (min)					
100	12	9	5	4	3	2
200	20	16	8	7	5	4
500	44	34	17	14	10	8
1000	75	58	30	24	18	13
2000	130	100	50	40	31	22
3000	175	134	67	55	42	30
4000	216	165	92	70	54	38
5000	250	195	95	82	65	45

From Schwab, Frevert, Edminster, and Barnes,
Soil and water conservation engineering, Wiley, New York.

To solve the equation we need to know each of the three factors on the right-hand side. The area A is measured by surveying, or from maps or aerial photographs.

To get the value of intensity I it is first necessary to estimate the gathering time of the catchment, that is, the longest time taken by surface run-off to get from any point in the catchment to the outlet. Table 3.1 gives values of gathering time for catchments of various size and slopes. Next we need to know the highest intensity of rain which is likely to last for the given gathering time. Local rainfall records should be used to estimate this if possible. When local records are not available an estimate can be made from Figure 3.1, which is derived from rainfall records in Australia and Africa. This figure shows the heaviest rainfall likely to occur on average once in 10 years. To get corresponding figures for shorter or longer periods we use the conversion factors of Table 3.2.

The constant C is a measure of the proportion of the rain which becomes run-off. On a corrugated-iron roof almost all the rain would run off, so C would be almost 1·0, while a well-drained sandy soil, where nine-tenths of the rain soaked in, would have a C value of 0·10. Table 3.3 gives some values of C. Where the catchment has several

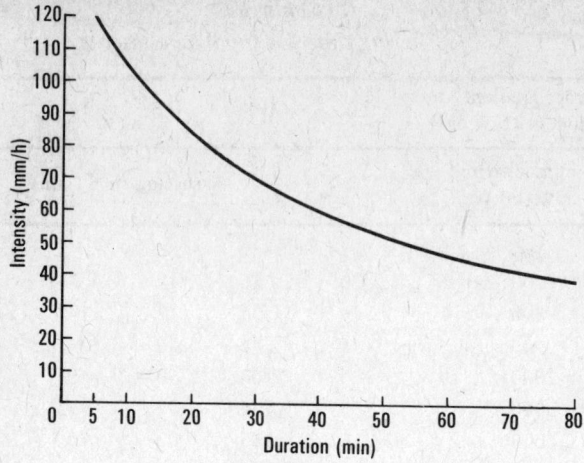

Figure 3.1.   The relationship between intensity of rainfall and duration.

Table 3.2
Rainfall probability conversion factors

2 years	0·90
5 years	0·95
10 years	1·00
25 years	1·25
50 years	1·50

different kinds of topography, the different values should be combined in proportion to the area of each.

3.1.2. Cook's method

This method was originally developed by an engineer of the US Soil Conservation Service. The method described here has been modified for tropical and sub-tropical countries, but the principle is the same, so the original name is still used.

First an assessment is made of the conditions in the catchment which most affect run-off. These are the kind of vegetation, the soil

Table 3.3
Values of run-off coefficient C

Topography and vegetation	Soil texture		
	Open sandy loam	Clay and silt loam	Tight clay
Woodland			
Flat 0-5 per cent slope	0·10	0·30	0·40
Rolling 5-10 per cent slope	0·25	0·35	0·50
Hilly 10-30 per cent slope	0·30	0·50	0·60
Pasture			
Flat	0·10	0·30	0·40
Rolling	0·16	0·36	0·55
Hilly	0·22	0·42	0·60
Cultivated			
Flat	0·30	0·50	0·60
Rolling	0·40	0·60	0·70
Hilly	0·52	0·72	0·82
Urban areas	30 per cent of area impervious	50 per cent of area impervious	70 per cent of area impervious
Flat	0·40	0·55	0·65
Rolling	0·50	0·65	0·80

From Schwab, Frevert, Edminster, and Barnes,
Soil and water conservation engineering, Wiley, New York.

permeability, and the slope. For each of these, the catchment condition is compared with the conditions listed in Table 3.4. The description is found in the table which most nearly fits the catchment, and the corresponding number is noted. Intermediate values can be used, for example, if half the catchment has heavy grass cover and the rest is not quite so dense, a value of 12 or 13 would be used. The arithmetic total of the number from each of the three columns is called the catchment characteristic (CC).

The area of the catchment is then measured, and using the area A and the catchment characteristic (CC), the maximum run-off can be read from Table 3.5. This gives the run-off for a 10-year probability,

Table 3.4
Catchment characteristics (CC)

Cover		Soil type and drainage		Slope	
Heavy grass	10	Deep, well drained soils	10	Very flat to gentle	5
Scrub or medium grass	15	Deep, moderately pervious soil	20	Moderate	10
Cultivated lands	20	Soils of fair permeability and depth	25	Rolling	15
Bare or eroded	25	Shallow soils with impeded drainage	30	Hilly or steep	20
		Medium heavy clays or rocky surfaces	40	Mountainous	25
		Impervious surfaces and waterlogged soils	50		

Select the most appropriate factor from each of these three lists and add them together.

Example

Heavy grass (10) on shallow soils with impeded drainage (30) and moderate slope (10):
CC = 10 + 30 + 10 = 50.

From Dept. of Conservation and Extension, Government of Rhodesia, *Handbook of basic instruction for dam construction.*

and the conversion factors given in Table 3.2 can be applied to get the corresponding figures for other time periods.

Another factor can be applied to take account of the shape of the catchment. Table 3.5 gives the run-off for a catchment which is roughly square or round. If the catchment is another shape the following conversion factors should be applied:

Square or round catchment	Long and narrow catchment	Broad and short catchment
1·0	0·8	1·25

3.2. Estimating water yield

In addition to knowing the probable rates of run-off, we need to know the total quantity which is likely to come from a catchment. The

Table 3.5
Run-off from small catchments

CC / A	25	30	35	40	45	50	55	60	65	70	75	80
5	0·2	0·3	0·4	0·5	0·7	0·9	1·1	1·3	1·5	1·7	1·9	2·1
10	0·3	0·5	0·7	0·9	1·1	1·4	1·7	2·0	2·4	2·8	3·2	3·7
15	0·5	0·8	1·1	1·4	1·7	2·0	2·4	2·9	3·4	4·0	4·6	5·2
20	0·6	1·0	1·4	1·8	2·2	2·7	3·2	3·8	4·4	5·1	5·8	6·5
30	0·8	1·3	1·8	2·3	2·9	3·6	4·4	5·3	6·3	7·3	8·4	9·5
40	1·1	1·5	2·1	2·8	3·5	4·5	5·5	6·6	7·8	9·1	10·5	12·3
50	1·2	1·8	2·5	3·5	4·6	5·8	7·1	8·5	10·0	11·6	13·3	15·1
75	1·6	2·4	3·6	4·9	6·3	8·0	9·9	11·9	14·0	16·4	18·9	21·7
100	1·8	3·2	4·7	6·4	8·3	10·4	12·7	15·4	18·2	21·2	24·5	28·0
150	2·1	4·1	6·3	8·8	11·6	14·7	18·2	21·8	25·6	29·9	35·0	40·6
200	2·8	5·5	8·4	11·7	15·3	19·1	23·3	28·0	33·1	38·5	45·0	52·5
250	3·5	6·5	9·7	13·2	17·2	21·7	27·0	32·9	39·6	46·9	55·0	63·7
300	4·2	7·0	10·5	14·7	19·6	25·2	31·5	38·5	46·2	54·6	63·7	73·5
350	4·9	8·4	12·6	17·2	23·2	30·2	37·8	46·3	53·8	62·5	71·5	81·0
400	5·6	10·0	14·4	19·4	25·6	33·6	42·2	51·0	60·0	69·3	79·5	90·0
450	6·3	10·5	15·5	21·5	28·5	36·5	45·5	55·5	65·5	76·0	86·5	97·5
500	7·0	11·0	17·0	23·5	31·0	40·5	51·0	62·0	73·0	84·0	95·0	106·5

From Dept. of Conservation and Extension, Government of Rhodesia, Dept. In-Service Manual.

A is the area of the catchment in hectares,
CC is the catchment characteristics from Table 3.4, and the run-off (in cubic metres per second is for a 10-year frequency.

To use this table in English units, let A be the catchment area in acres, and the figure in the table multiplied by 14·3 will give the run-off in cubic feet per second.

total annual run-off is called the *water yield*, although we may be more interested in shorter periods, like the monthly average flow or the amount from individual storms. The usefulness of water for irrigation or domestic supplies does not only depend upon the total amount, but also upon when it is available and how reliable the supply will be. The average flow might be misleading if we do not know the likely variation on either side of the average and the probable minimum flow. The design of an irrigation scheme using a constant reliable flow would be very different from a scheme which requires storage to even out an unreliable and varying flow. Estimates of surface water therefore depend very much on having rainfall and streamflow records, and the longer and more reliable the records, the more accurate is the estimate based on them.

The methods of estimating water yield are quite different in arid climates and in humid climates. In humid climates the water table is most of the time fairly close to the surface, and higher than the bed of streams and rivers. There is therefore a steady seepage of ground water into the streams, in addition to the direct run-off from storms. It is not possible to tell which of the water in a stream has come from seepage flow and which from storm flow, and so the total flow can not be correlated with rainfall records, and the only way to predict water yield is from past records of flow.

In countries with sufficient data on streamflow, maps can be drawn showing *isopleths*, or lines of equal run-off. Naturally these resemble the rainfall maps, but the proportion of run-off is greater when the total rainfall is more, so the difference between low and high rainfall is magnified in run-off maps.

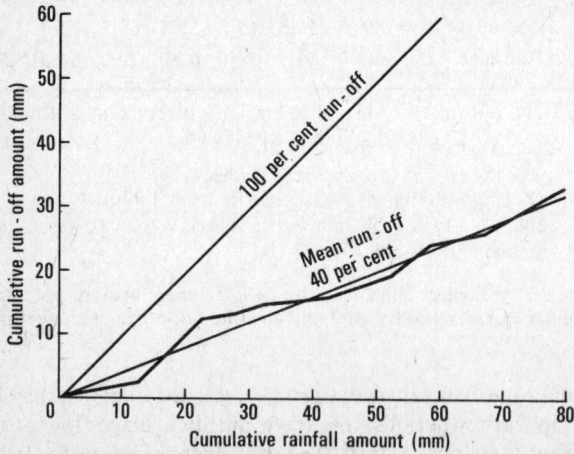

Figure 3.2.   The average run-off in arid or semi-arid climates can be found by plotting cumulative totals of measured rainfall against measured run-off.

In arid regions there is no reservoir of groundwater, and so no seepage flow. The yield therefore consists entirely of storm run-off and this can be estimated from rainfall records.

The amount of run-off is the rainfall minus the losses, that is,

Q (run-off) = P (rainfall) $- L$ (losses).

In semi-arid climates this method can be used to estimate annual run-off from annual rainfall by subtracting the estimated annual evapotranspiration. This is a function of land use and latitude, and can vary from 300 mm to 800 mm (1 - 2·5 ft) per year. If the cumulative run-off from a catchment is plotted against cumulative rainfall, the average losses can be determined from the slope of the graph as in Figure 3.2. Such a plot can be made using daily or weekly data or from individual storms.

In arid climates the loss is the infiltration, and an estimate of yield can be obtained by applying the formula to each storm, assuming the same loss from each storm, with values from 10 mm to 20 mm (0·4 - 0·8 in) per storm.

A more accurate method is to recognize that the losses are going to vary according to the amount of rainfall in the storm, and according to the amount of moisture which can be absorbed by the soil. This is the basis of the formula of the U.S. Soil Conservation Service, which in metric units is

$$Q = \frac{(I - 0 \cdot 2S)^2}{I + 0 \cdot 8S} \quad ,$$

Table 3.6
Values of S (mm) for water yield formula

Soil type	Number of days since last storm which caused run-off		
	More than 5	2 - 5	Less than 2
Good permeability, for example, deep sands	150	75	50
Medium permeability, for example, sandy clay loams and clay loams	100	50	25
Low permeability, for example, clays	50	25	25

Intermediate values may be used.
From U.S.D.A. Soil Conservation Service Manual, Engineering Handbook.

where Q is the run-off in millimetres,

I is the storm rainfall in millimetres,

S is the greatest amount of rainfall in millimetres which can soak into the soil during the storm.

One possibility is to assume a constant value of S for a given catchment. More accurate estimates can be made by assuming that if storms occur in quick succession the soil will not have time to dry out in between. Table 3.6 shows some values of S which make allowance for this and for the different storage capacity of different soils. Figure 3.3 shows the equation plotted for various values of S. In English units the formula is the same with all items in inches.

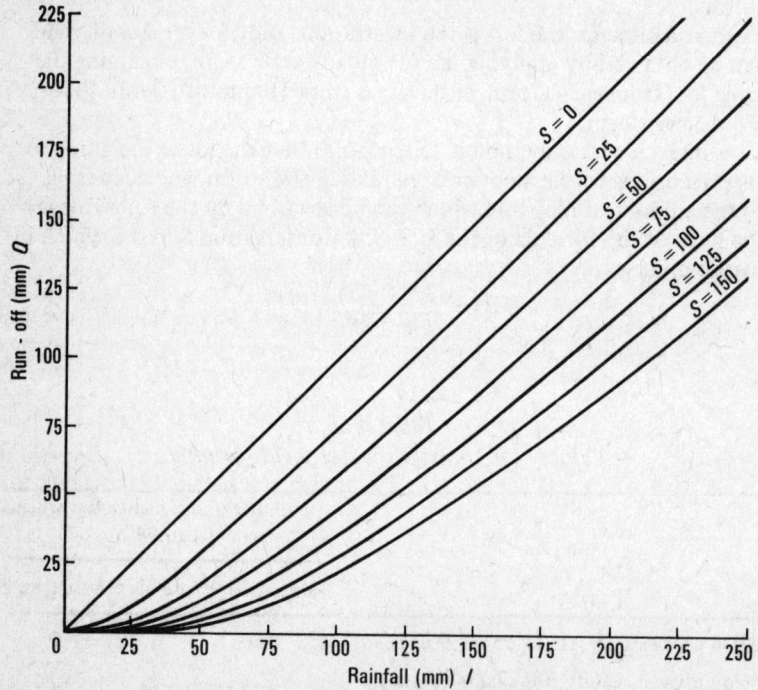

Figure 3.3.   The relationship between run-off Q mm, and rainfall I mm, for various values of catchment storage S mm, using the formula

$$Q = \frac{(I - 0 \cdot 2S)^2}{(I + 0 \cdot 8S)}.$$

Values of S are shown in Table 3.6.

3.3. Measuring water

3.3.1. Stream and river gauging

Most methods of measuring the flow of streams or rivers depend **upon** measuring the average velocity and the cross-sectional area and calculating the flow from

Q (m^3/s) $= A$ (m^2) $\times V$ (m/s)

or Q (ft^3/s) $= A$ (ft^2) $\times V$ (ft/s).

The fps unit ft^3/s is commonly referred to in English usage as the *cusec*. The American abbreviation is cfs. The metric unit m^3/s is similarly called the *cumec*. Because m^3/s is a large unit, smaller flows are measured in litres per second (l/s).

The simplest way to estimate the velocity is to measure the time taken for some small floating object to travel a measured distance downstream. The velocity is not the same at all places in the stream, being slower at the sides and edges and faster on the surface, as shown in Figure 3.4. Taking 0·8 of the surface velocity as measured by the float gives an approximate value for the average velocity.

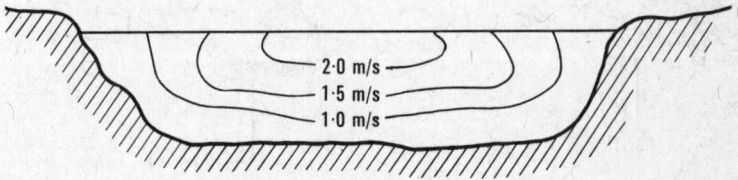

Figure 3.4.   A typical distribution of velocity of flow in a stream or open channel.

Alternatively, the velocity can be measured below the surface by attaching a submerged weight to a float. The float and weight move downstream together at the velocity of the stream at the depth where the weight is suspended. At about half the stream depth the velocity is approximately the same as the average velocity for the whole stream. Float methods are only suitable for straight streams or canals where the flow is fairly even and regular.

Another method is to pour into the stream a quantity of strongly coloured dye, and to measure the time for this to flow a known distance. The dye should be added quickly and at a constant rate with a sharp cut-off, so that it travels downstream in a cloud. The time is measured for the first and last of the dye to reach the downstream point and an average of the two times is used to calculate the average velocity.

In turbulent streams the cloud of dye is dispersed and cannot be seen and measured, so a salt solution is used and detected by chemical analysis. There are two ways of using salt solutions. One is to add the solution as a single application, like the dye, and find when it reaches the downstream sampling point by analysing samples of the water taken at regular intervals. The other is the dilution method. A salt solution of known concentration is added at a measured constant rate, and a sample is taken somewhere downstream after it will have been thoroughly mixed. The amount which the salt is diluted will depend on the rate of flow. We know the strength of the solution and the rate at which it is added, and we measure the diluted concentration in the river, so from this we can calculate the flow in the river.

(a)   (b)

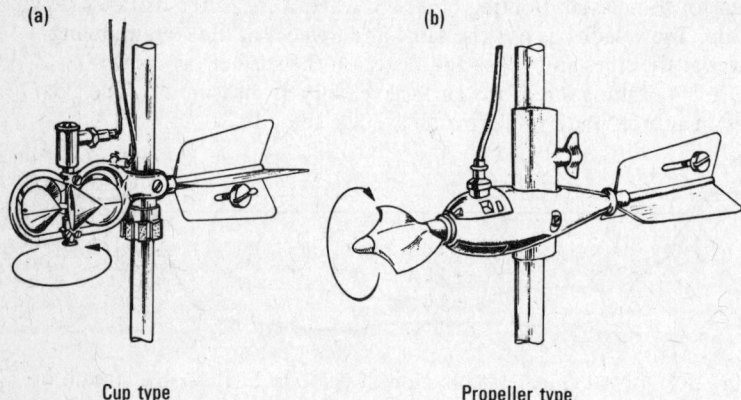

Cup type   Propeller type

Figure 3.5.   Current meters used to measure the velocity of flow in streams, rivers, or canals.

More accurate determination of velocity is obtained by using a current meter. The two main types are illustrated in Figure 3.5. The conical cup type revolves about a vertical axis, and the propeller type about a horizontal axis. In each case the speed of revolution is proportional to the water velocity, and the number of revolutions in a given time is counted, either on a digital counter or as clicks heard in earphones worn by an operator. In shallow streams small current meters will be mounted on rods and held by wading operators. When measurements of the maximum flood flows have to be measured on big rivers an overhead cableway is installed, well above maximum flood level, and the current meter is lowered on cables into the river, with weights to

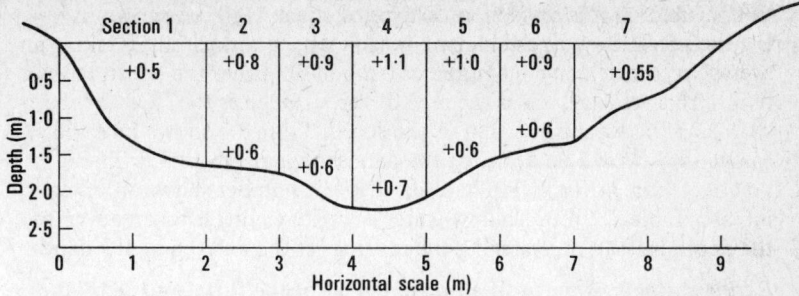

Figure 3.6.   Estimating the flow in a stream from measurements with a current meter. The calculations for this example are shown in Table 3.7.

Table 3.7
Calculation of streamflow from current-meter readings

1	2	3	4	5	6	7	8
Section	Flow velocity (m/s)			Depth (m)	Width (m)	Area (m^2) 5 x 6	Flow (m^3/s) 4 x 7
	0·2D	0·8D	Mean				
1	-	-	0·5	1·3	2·0	2·6	1·30
2	0·8	0·6	0·7	1·7	1·0	1·7	1·19
3	0·9	0·6	0·75	2·0	1·0	2·0	1·50
4	1·1	0·7	0·9	2·2	1·0	2·2	1·98
5	1·0	0·6	0·8	1·8	1·0	1·8	1·44
6	0·9	0·6	0·75	1·4	1·0	1·4	1·05
7	-	-	0·55	0·7	2·0	1·4	0·77
							Total 9·23

D is the depth of the stream at the mid-point of the section. See Figure 3.6.

hold it against the river flow. Current meters can also be used from bridges.

A current meter measures the velocity at a single point, and several measurements are required to calculate the total flow. The procedure is to measure and plot on graph paper the cross-section of the stream and to imagine that it is divided into strips of equal width, as shown in Figure 3.6. The average velocity for each strip is estimated by taking an average of the velocity measured at 0·2 and 0·8 times the depth at that point. This velocity, times the area of the strip, gives the flow for the strip, and the total is the sum of the strips. Table 3.7 shows how the calculations would be done for the data shown in Figure 3.6. In practice, more strips would be used than the number shown in Figure 3.6 and Table 3.7. For shallow water a single reading is taken at 0·6 of the depth instead of averaging the readings at 0·2 and 0·8 of the depth.

Rating a gauging station. If a measurement of the flow is made by the current-meter method on different occasions when the river is flowing at many different depths, these measurements can be used to draw a graph of amount of flow against depth of flow. The depth of flow of a river is called *stage*, and when a curve has been obtained for discharge against stage, the gauging station is described as rated. Subsequent estimates of flow can be obtained by measuring the stage at a fixed gauging post, and reading off the flow from the rating curve. If the stream cross-section changes through erosion or deposition a new rating curve has to be drawn up. For accuracy on large rivers two separate curves are used, one when the river is rising and another when it is falling. When establishing the rating curve, each gauging can only be done when the river happens to be at the right stage, so when gauging stations are in remote areas it can take many years work.

3.3.2. Gauging weirs and flumes

The calculation of flow from measured velocity is laborious, and is simplified by passing the flow through some constructed channel where the flow can be deduced by measuring the depth of flow. This is the principle of a large variety of devices for measuring water flow.

Notched weirs; 90° V-notch. A simple and accurate method is the 90° V-notch shown in Figure 3.7. The water must fall freely over the downstream side. When the water on the downstream side is so high as to prevent free fall over the weir this is called a drowned or submerged weir. Some kinds of weir will work when submerged or partly submerged, but not a V-notch.

On the upstream side there should be a stilling basin where turbulence is evened out so that the water approaches the notch smoothly. The notch can be made of metal, or it can be a metal edge set in a brick or concrete wall, but the notch must have a sharp edge. It must be upright and level.

(a)

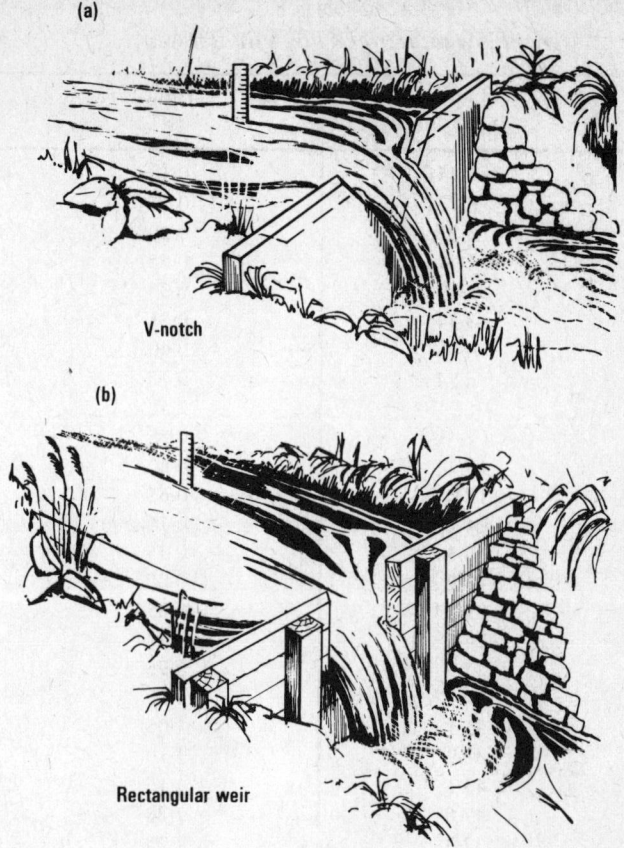

V-notch

(b)

Rectangular weir

Figure 3.7.   Measuring stream flow with sharp-crested weirs.
(a) 90° V-notch weir. (b) Rectangular notch weir.

To read the depth of flow through the notch a measuring scale is set
in the stilling pool in a position where it can be easily read. The zero of
the scale is set level with the lowest point of the notch. The scale should
be set well back from the notch so that it is not affected by the draw-
down curve shown in Figure 3.7.

V-notch weirs are portable and simple to instal in either temporary
or permanent positions. They are particularly suitable for small flows,

Table 3.8
Flow rates over a 90° V-notch weir

Head (mm)	Flow (l/s)	Head (ft)	Flow (ft³/s)
40	0·441	0·10	0·009
50	0·731	0·15	0·023
60	1·21	0·20	0·046
70	1·79	0·25	0·080
80	2·49	0·30	0·125
90	3·34	0·35	0·184
100	4·36	0·40	0·256
110	5·54	0·45	0·343
120	6·91	0·50	0·445
130	8·41	0·55	0·564
140	10·2	0·60	0·70
150	12·0	0·65	0·85
160	14·1	0·70	1·03
170	16·4	0·75	1·22
180	18·9	0·80	1·43
190	21·7	0·85	1·66
200	24·7	0·90	1·92
210	27·9	0·95	2·19
220	31·3	1·00	2·49
230	35·1	1·05	2·81
240	38·9	1·10	3·15
250	43·1	1·15	3·52
260	47·6	1·20	3·91
270	52·3	1·25	4·33
280	57·3		
290	62·5		
300	68·0		
350	100·0		

From U.S.D.I. Bureau of Reclamation, *Water measurement manual.*

and the only drawback is that they must have a free fall of more than the depth of flow, so they are not suitable for use in irrigation canals.

Table 3.8 gives discharge values through small 90° V-notch weirs.

For larger flows the *rectangular weir* (Figure 3.7) is more suitable because the width can be chosen so that it can pass the expected flow

Table 3.9
Flow rates over a rectangular weir (with end contractions)

Head (mm)	Flow (l/s per metre of crest length)	Head (ft)	Flow (ft³/s per foot of crest length)
30	9·5	0·10	0·103
40	14·6	0·15	0·188
50	20·4	0·20	0·286
60	26·7	0·25	0·395
70	33·6	0·30	0·514
80	40·9	0·35	0·658
90	48·9	0·40	0·801
100	57·0	0·45	0·950
110	65·6	0·50	1·11
120	74·7	0·55	1·28
130	84·0	0·60	1·44
140	93·7	0·65	1·62
150	103·8	0·70	1·86
160	114·0	0·75	2·06
170	124·5	0·80	2·25
180	136·0	0·85	2·46
190	146·0	0·90	2·68
200	158·5	0·95	2·91
210	169·5	1·00	3·13
220	181·5	1·05	3·30
230	193·5	1·10	3·53
240	205·5	1·15	3·76
250	218·5	1·20	4·00
260	231·0	1·25	4·23
270	244·0	1·30	4·47
280	257·5	1·35	4·69
290	271·0	1·40	4·96
300	284·0	1·45	5·23
310	298·0	1·50	5·47
320	311·5		
330	326·0		
340	340·0		
350	354·0		
360	368·5		
370	383·5		
380	398·0		

From U.S.D.I. Bureau of Reclamation, *Water measurement manual.*

at a suitable depth. Table 3.9 gives the discharge for various sizes of rectangular notch.

Flumes. If the flow in a channel is squeezed through a narrower section of channel it flows faster, but there is a *critical velocity* which can not be exceeded. If the water is forced to flow at critical velocity the rate of flow can be accurately determined by measuring the depth of flow. Several flumes have been designed on this principle and one of the best known is the *Parshall flume* (Figure 3.8). This can be built in a large variety of sizes capable of measuring up to 90 m³/s (3200 ft³/s), and can be constructed from a wide variety of materials such as wood, concrete, metal, or combinations of these. It can be prefabricated in a workshop, or can be cast in concrete on site using re-usable form-work. In short it is very versatile and can be used in most circumstances.

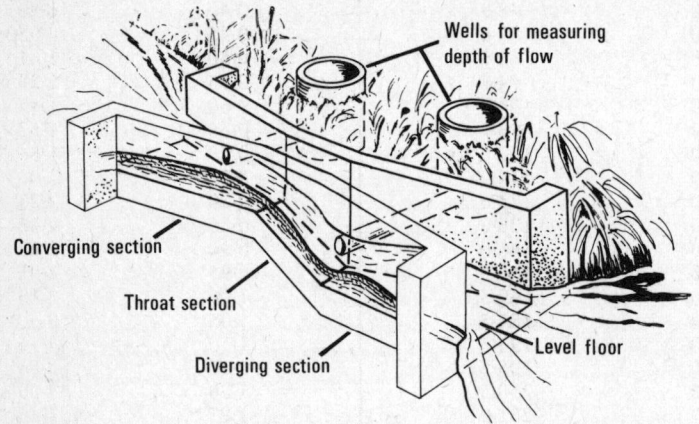

Figure 3.8.   Parshall measuring flume (re-drawn from California Agricultural Experimental Station Circular 473).

Only a small head is lost through the flume, and this makes it useful for measuring the flow in irrigation canals. The head required can be even further reduced by operating the flume in the submerged condition but this requires the use of a second gauging scale at the downstream end, and calculating a correction factor according to the amount of submergence.

The rates of flow in Parshall flumes of small and medium size are shown in Table 3.10.

Another critical-depth flume of a design rather similar to the Parshall is the *Washington trapezoidal flume*, which is particularly useful as a **portable** flume for spot measurements of small flows in unlined

Table 3.10
Flow rates in Parshall flumes

(a) *Metric units*

Throat width (mm)	76·2	152·4	228·6	304·8
Head (mm)	Flow (m³/s)			
30	0·0008	0·0015	0·0025	0·0033
40	0·0012	0·0024	0·0039	0·0052
50	0·0017	0·0034	0·0055	0·0073
60	0·0023	0·0045	0·0072	0·0096
70	0·0029	0·0057	0·0091	0·0121
80	0·0035	0·0070	0·0112	0·0149
90	0·0042	0·0085	0·0134	0·0178
100	0·0050	0·0100	0·0158	0·0209
110	0·0058	0·0117	0·0183	0·0241
120	0·0066	0·0134	0·0209	0·0275
130	0·0075	0·0152	0·0236	0·0311
140	0·0084	0·0171	0·0264	0·0348
150	0·0094	0·0190	0·0294	0·0386
160	0·0103	0·0211	0·0324	0·0426
170	0·0114	0·0232	0·0356	0·0467
180	0·0124	0·0254	0·0388	0·0510
190	0·0135	0·0276	0·0422	0·0554
200	0·0146	0·0300	0·0456	0·0598
225	0·0175	0·0361	0·0546	0·0716
250	0·0206	0·0426	0·0642	0·0840
275	0·0239	0·0496	0·0742	0·0971
300	0·0274	0·0569	0·0848	0·1108
325	0·0310	0·0646	0·0958	0·1252
350	0·0348	0·0726	0·1073	0·1401

irrigation furrows. It can be prefabricated in fibre glass or thin sheet-metal and installed in a few minutes. Dimensions are shown in Figure 3.9 and the rating in Table 3.11.

A group of special-purpose flumes called *H flumes* were designed by the U.S. Soil Conservation Service for measuring flows accurately and continuously from run-off plots or small experimental catchments. The design requirements were that the flume should measure low flows

(b) *English units*

Throat width (in)	3	6	9	12
Head (ft)	Flow (ft^3/s)			
0·10	0·03	0·05	0·09	0·11
0·15	0·05	0·10	0·17	0·20
0·20	0·08	0·16	0·26	0·35
0·25	0·12	0·23	0·37	0·49
0·30	0·15	0·31	0·49	0·64
0·35	0·20	0·39	0·62	0·80
0·40	0·24	0·48	0·76	0·99
0·45	0·29	0·58	0·90	1·19
0·50	0·34	0·69	1·06	1·39
0·55	0·39	0·80	1·23	1·62
0·60	0·45	0·92	1·40	1·84
0·65	0·51	1·04	1·59	2·08
0·70	0·57	1·17	1·78	2·33
0·75	-	1·31	1·98	2·58
0·80	-	1·45	2·18	2·85
0·85	-	1·59	2·39	3·12
0·90	-	1·74	2·61	3·41
0·95	-	1·90	2·84	3·70
1·00	-	2·06	3·07	4·00
1·05	-	2·22	3·31	4·31
1·10	-	2·40	3·55	4·62
1·15	-	2·57	3·80	4·94
1·20	-	2·75	4·06	5·28

From U.S.D.I. Bureau of Reclamation, *Water measurement manual.*

accurately but still have a good capacity for high flows, and not need a stilling pond. It also had to be able to pass run-off containing a heavy silt load. When manufactured accurately to the given dimensions (Figure 3.10) the flume does not need calibration, and the rating can be taken from Table 3.12. This flume can operate partially submerged, and a simple correction curve is applied (Figure 3.11). A small range (Type HS) is suitable for small flows up to 23 l/s (0·8 ft^3/s), and a larger range (Type HL) for flows up to 3 m^3/s (115 ft^3/s).

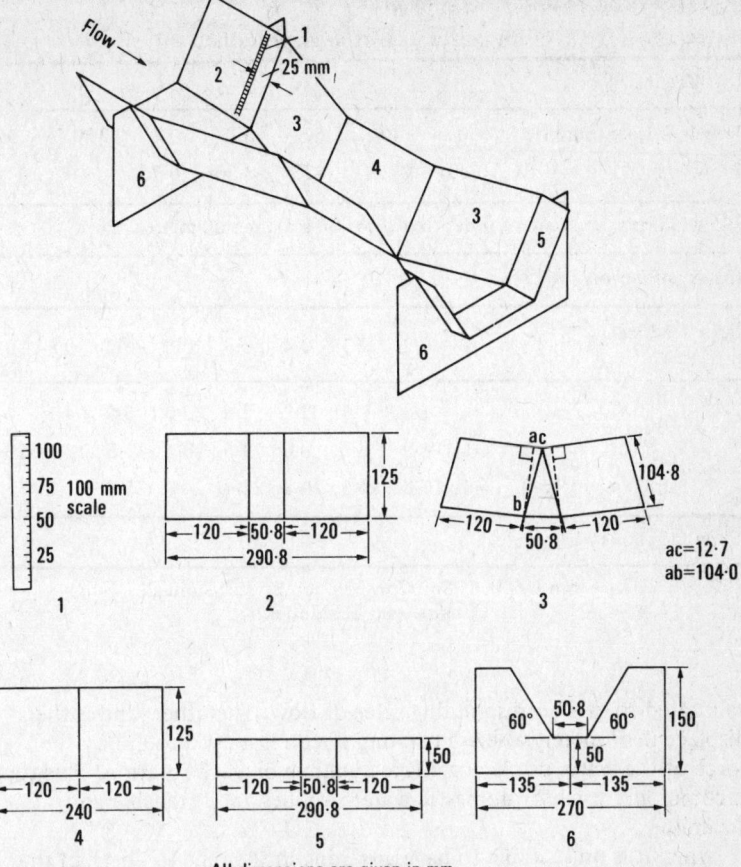

Figure 3.9.   The Washington flume. Dimensions in millimetres.
(Metric conversion of details from the U.S.D.A. Soil Conservation
Service National Engineering Handbook, Section 15, Chapter 9.)

3.3.3. Measurement of pipe flow

Flow in pipes is measured either directly by a meter, or calculated
from measurements of pressure.

Impeller-type meters are like the current meters used for stream
gauging, that is the meter counts the number of revolutions and this is

Table 3.11
Flow rates in Washington flumes

(a) *Metric units*

Depth of flow (mm)	30	40	50	60	70	80	90
Flow (l/s)	0·10	0·20	0·33	0·50	0·75	1·07	1·43

Flow in litres per second for depth of flow on scale in millimetres.

(b) *English units*

Scale reading (in)	0	0·1	0·2	0·3	0·4	0·5	0·6	0·7	0·8	0·9
1			1·5	1·8	2·2	2·6	3·0	3·5	4·1	4·7
2		5·3	6·1	6·9	7·7	8·6	9·5	10·4	11·5	12·7	13·8
3	15·0	16·4	17·8	19·3	20·8	22·4	24·0	25·8		

Flow in gallons per minute.

From U.S.D.A. Soil Conservation Service Manual,
Engineering Handbook.

calibrated to give corresponding rates of flow. The other kind is the displacement meter, where a rotating piston is moved round by the force of the water displacing a fixed volume on each rotation. This type is commonly used for domestic water supplies and is reliable and accurate.

When it is undesirable to have anything in the pipe to obstruct the flow, a *venturi meter* is used (Figure 3.12). A slight constriction in the pipe speeds up the flow and the pressure is measured both at the constriction and at the normal pipe diameter. From a comparison of the two pressures the flow rate can be obtained from tables. The *Dall tube* is a commercial form of venturi meter, and can be used in almost any size.

The flow in pipes may also be measured at the point where it discharges from the pipes. The simplest way is direct volumetric measurement, that is, we measure the time to fill a container of known size or we use a graduated container to measure the amount collected in a fixed time.

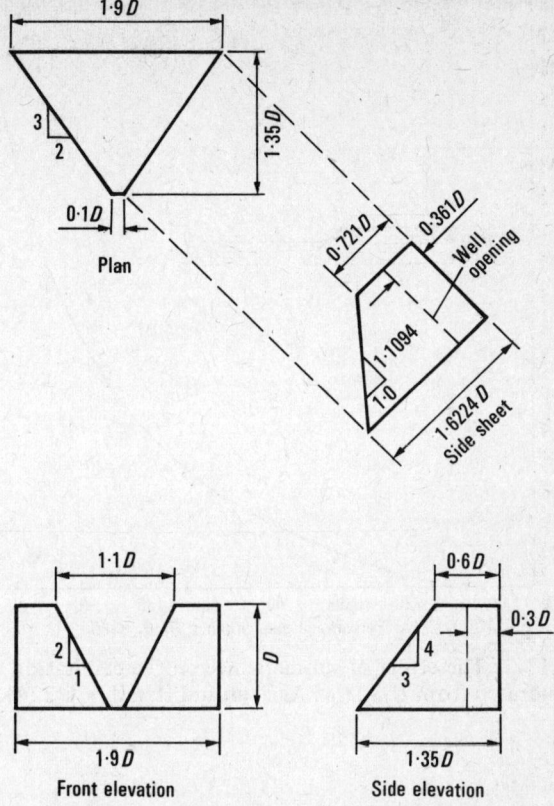

Figure 3.10.   The proportions of the H-flume (re-drawn from U.S.D.A. Agricultural Research Service Agricultural Handbook 224).

The jet of water discharging from a pipe under pressure depends upon the rate of flow in the pipe, and the rate can be calculated by measuring the jet. If the pipe can be arranged to discharge vertically upwards we measure the height to which it rises above the end of the pipe and calculate the rate of flow from the appropriate formula shown in Figure 3.13. Estimates of discharge can also be made from measurements of the trajectory from horizontal and sloping pipes.

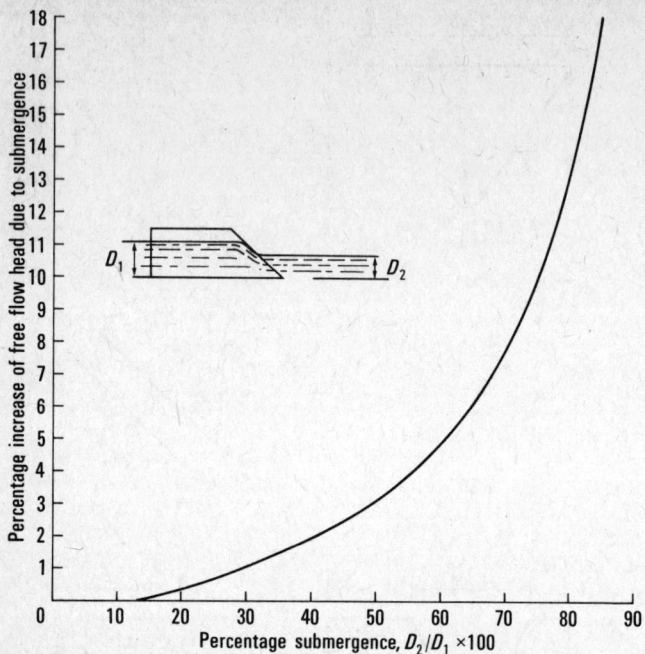

Figure 3.11. The effect of submergence on the calibration of an H-flume (re-drawn from U.S.D.A. Agricultural Handbook 224).

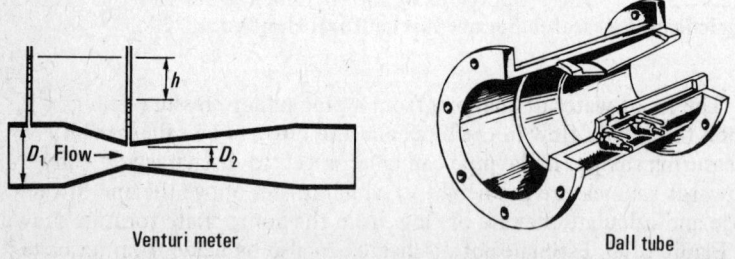

Venturi meter

Dall tube

Figure 3.12. The venturi tube or venturi meter, used to measure flow in pipes.

Table 3.12
Flow rates in H flumes

(a) Metric units

Size of flume (maximum depth in millimetres)	Flow in litres per second for depth of flow in millimetres						
	25	50	100	250	500	750	1000
250	0·45	1·42	3·41	32·9	-	-	-
500	0·60	1·95	4·52	38·2	186·1	-	-
750	0·75	2·82	5·85	42·1	201·1	526·8	-
1000	0·93	3·77	8·48	48·1	208·2	546·8	1090·3

(b) English units

Size of flume (maximum depth in feet)	Flow in cubic feet per second for depth of flow in feet							
	0·1	0·4	0·8	1·0	1·5	2·0	2·5	3·0
0·5	0·0101	0·204						
1·0	0·0150	0·244	1·16	1·96				
1·5	0·0200	0·283	1·27	2·09	5·41			
2·0	0·0248	0·323	1·38	2·25	5·65	11·1		
2·5	0·0298	0·363	1·49	2·41	5·91	11·5	19·4	
3·0	0·0347	0·403	1·61	2·57	6·24	11·9	19·9	31·0

From U.S.D.A. Agricultural Handbook 224.

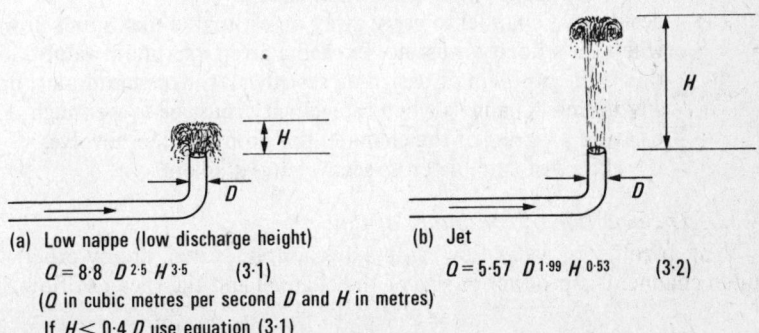

(a) Low nappe (low discharge height)

$$Q = 8 \cdot 8 \ D^{2 \cdot 5} \ H^{3 \cdot 5} \qquad (3 \cdot 1)$$

(*Q* in cubic metres per second *D* and *H* in metres)

If $H < 0 \cdot 4 \ D$ use equation (3·1)

If $H > 1 \cdot 4 \ D$ use equation (3·2)

If $0 \cdot 4D < H < 1 \cdot 4D$ calculate both equations and take the average.

(b) Jet

$$Q = 5 \cdot 57 \ D^{1 \cdot 99} \ H^{0 \cdot 53} \qquad (3 \cdot 2)$$

Figure 3.13.   Calculation of flow in pipes from the height to which the discharge rises

4. Water flow in channels and pipes

The conveyance of water in open channels and in pipes is important in many branches of agricultural development, such as irrigation, drainage, soil conservation, and domestic water supplies, and each of these has its own special requirements. However, the principles of designing the pipes and channels are always the same, so we can consider these in general terms before going on to the particular applications.

4.1. Open channel flow

The same principles are used to solve each of the most frequently encountered problems of channel flow, which are:
- (1) estimating the flow in a channel when the cross-section, gradient, depth, etc. are known or can be measured. This arises in irrigation canals, ditches, and natural watercourses;
- (2) estimating the depth of flow at which a given channel will carry a given rate of flow—this is the problem of forecasting how high a river flood will rise, or how deep the flow will be in an irrigation canal;
- (3) designing a channel which will carry away a given rate of flow as rapidly as possible—this arises when storm run-off has to be lead away from buildings or structures;
- (4) designing a channel to carry away an estimated maximum flow, when the velocity must not exceed a given maximum value— this is the problem of designing stormwater diversion drains, or any unlined channels when the velocity must be low enough to avoid scouring of the channel; the problem then involves the choice of suitable cross-section and gradient.

4.1.1. The equation of continuity of flow

The quantity of water flowing in a ditch, drain, canal, or any other open channel is a product of size of the channel and the speed of flow,

$$Q = A \times V,$$

where Q is the flow in cubic metres per second,
$\quad A$ is the cross-sectional area of the channel in square metres, and
$\quad V$ is the average velocity of flow in metres per second.

In English units:
$$Q \text{ (ft}^3/\text{s)} = A \text{ (ft}^2) \times V \text{ (ft/s)}.$$

At any cross-section in the channel the water does not all flow at the same speed. When considering the measurement of flow in streams and rivers in Chapter 3 we saw that the water at the sides and bottom is slowed down by frictional resistance, while the water in the middle flows faster. In artificial channels, such as concrete-lined irrigation canals, the slowing down at the sides is less than in the case of vegetated channels, but in all cases we can use a theoretical average velocity for the whole flow.

The equation is sometimes called the equation of continuity of flow. In steady flow, that is, when the discharge Q is constant, if there is a change in the cross-sectional area of the channel, there must be a corresponding change in velocity. If the stream channel narrows, the velocity increases so that
$$Q = A_1 \times V_1 = A_2 \times V_2 = A_3 \times V_3.$$

There can also be two different kinds of flow which both have the same discharge, in the same shape of channel, but at different depths and different velocities. For example, a rectangular channel 10 m wide could carry 20 m^2/s at a depth of 2 m and velocity of 1 m/s, or at a depth of 0·5 m and a velocity of 4 m/s:
$$Q = 20 = 10 \times 2 \times 1 = 10 \times 0·5 \times 4.$$
$$A_1 \times V_1 = A_2 \times V_2.$$

The fast, shallow flow is called *super-critical flow, low-stage flow,* or *shooting flow,* and the slower, deeper flow is called *sub-critical flow,* or *high-stage flow.* A good example of the two kinds is seen at dam spillways, where the water comes down the spillway in fast super-critical flow then there is a standing wave or hydraulic jump at the bottom of the spillway then the water continues downstream in low-velocity, deeper flow.

Apart from spillways and chutes we tend to avoid the use of super-critical velocities because flow in this condition is unstable, and an obstruction or variation in the channel can cause a sudden change to sub-critical flow which could overtop the channel. Also the high velocities of super-critical flow cause erosion of channels if they are not lines with concrete.

4.1.2. *Velocity of flow*

The velocity of water flowing in an open channel is affected by the following factors.

Gradient or slope. Velocity increases when the gradient is steeper.

Roughness. The contact between the water and the channel exercises a frictional resistance which depends on the smoothness or roughness of the channel. A concrete-lined canal has less resistance than one whose sides are choked with vegetation. It may be desirable to design a smooth channel when it is required to pass as much water as possible, or to increase the roughness when it is desired to slow the water down to non-scouring velocity.

Shape. Channels can have the same cross-sectional area, gradient, and roughness, but still have different velocities according to their shape. The reason is that water close to the sides and bottom of the channel is slowed by the friction effect, so a channel whose shape provides least area of contact with the water will have least frictional resistance and so a greater velocity. The parameter used to measure this effect of shape is called the *hydraulic radius* of the channel. It is defined as the cross-sectional area divided by the wetted perimeter, which is the length of the cross-section of the channel which is in contact with the water. Hydraulic radius thus has units of length, and it may be represented by either M or R. It is also sometimes called *hydraulic mean radius* or *hydraulic mean depth*. Figure 4.1 shows how channels can

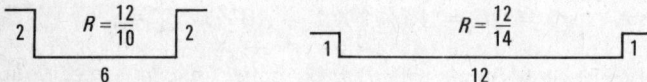

Figure 4.1.   Channels with the same cross-sectional area can have different hydraulic radius.

have the same cross-sectional area but different hydraulic radius. If all other factors are constant, then the lower the value of R, the lower will be the velocity. Earth channels are usually given a cross-section which is either trapezoidal or a smooth surve which approximates to a part of a parabola. Channels originally excavated as trapezoidal sections tend towards a parabolic shape in time. V-shaped sections are undesirable for earth channels because the lowest point is liable to scour. The hydraulic radius and other dimensions of the most common channel shape are shown in Figure 4.2.

All these variables which affect velocity of flow have been brought together in a very useful empirical equation called the *Manning formula*. In metric units this formula is

$$V = \frac{R^{2/3} \times S^{1/2}}{n} ,$$

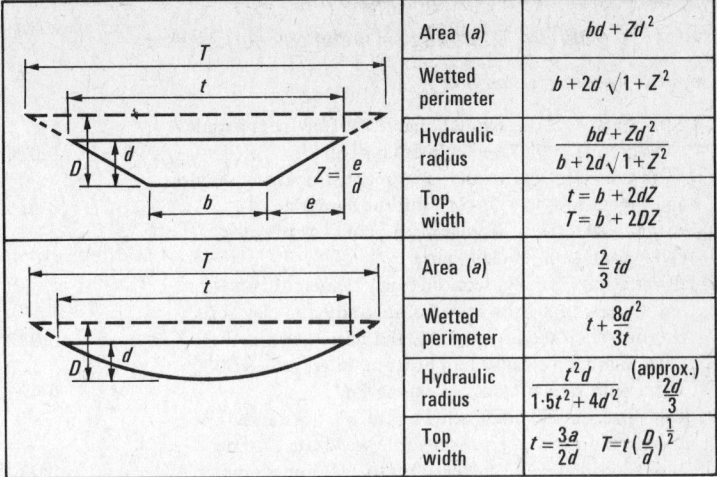

	Area (*a*)	$bd + Zd^2$
	Wetted perimeter	$b + 2d\sqrt{1+Z^2}$
	Hydraulic radius	$\dfrac{bd + Zd^2}{b + 2d\sqrt{1+Z^2}}$
	Top width	$t = b + 2dZ$ $T = b + 2DZ$
	Area (*a*)	$\dfrac{2}{3}td$
	Wetted perimeter	$t + \dfrac{8d^2}{3t}$
	Hydraulic radius	$\dfrac{t^2 d}{1 \cdot 5t^2 + 4d^2}$ (approx.) $\dfrac{2d}{3}$
	Top width	$t = \dfrac{3a}{2d}$ $T = t\left(\dfrac{D}{d}\right)^{\frac{1}{2}}$

Figure 4.2.   Basic dimensions of common channel sections.

where *V* is in metres per second,
   R is in metres,
   S is in metres per metre, and
   n is a coefficient, known as Manning's *n*, or Manning's roughness coefficient.
Some values for channel flow are listed in Table 4.1.
   In English units the formula is usually given in the form

$$V = \frac{1 \cdot 486 \times R^{2/3} \times S^{1/2}}{n},$$

where *V* is the average velocity of flow in feet per second,
   R is the hydraulic radius in feet,
   S is the gradient in feet per foot, and
   n is the roughness coefficient which has the same values as in the metric form of the equation.
The figure $1 \cdot 486$ arises because the formula was originally derived in metric units and this is the cube root of $3 \cdot 28$, the number of feet in one metre. The value of $1 \cdot 486$ is more precise than necessary. Considering the accuracy of the values of *n* it would be more appropriate to use $1 \cdot 5$ instead of $1 \cdot 486$.
   If any three of the four variables in the Manning equation are known, the fourth one can be found. If we can measure or estimate the values of *R*, *S*, and *n* in an existing channel, we can calculate the velocity.

Table 4.1
Values of Manning's roughness coefficient n

(*a*) *Channels free from vegetation*   *n*

Uniform cross-section, regular alignment free from pebbles and vegetation, in fine sedimentary soils	0·016
Uniform cross-section, regular alignment, free from pebbles and vegetation, in stiff clay soils or hardpan	0·018
Uniform cross-section, regular alignment, few pebbles, little vegetation, in clay loam	0·020
Small variations in cross-section, fairly regular alignment, few stones, thin grass at edges, in sandy and clay soils, also newly cleaned, ploughed, and harrowed channels	0·0225
Irregular alignment, ripples on bottom, in gravelly soil or shale, with jagged banks or vegetation	0·025
Irregular section and alignment, scattered rocks and loose gravel on bottom, or considerable weeds on sloping banks, or in gravelly material up to 150 mm diameter	0·030
Eroded irregular channels, channels blasted in rock	0·030

(*b*) *Vegetated channels*

Short grass (50-150 mm)	0·030-0·060
Medium grass (150-250 mm)	0·030-0·085
Long grass (250-600 mm)	0·040-0·150

(*c*) *Natural stream channels*

Clean and straight	0·025-0·030
Winding, with pools and shoals	0·033-0·040
Very weedy, winding, and overgrown	0·075-0·150

When using the equation as an aid to the design of channels we can choose a velocity, gradient, and roughness and see what hydraulic radius is required. Or we can assume values of velocity, shape, and roughness and find the gradient. Other design applications are discussed in the next section.

Another useful empirical formula for estimating the velocity of streams or canals is *Elliot's open-ditch formula*, which uses English units:

$$V = \sqrt{\left(R \times \frac{3h}{2}\right)},$$

where V is the average velocity in feet per second,
  R is the hydraulic radius in feet, and
  h is the average gradient in feet per mile.
The equation assumes a value for Manning's n of 0·02 and so is suitable for clean, straight, natural streams, or unlined irrigation canals, or drainage ditches clear of weeds.

4.2. The design of open channels

Practical problems of designing open channels may vary a lot in detail, but the principle is usually the same. The designer has some

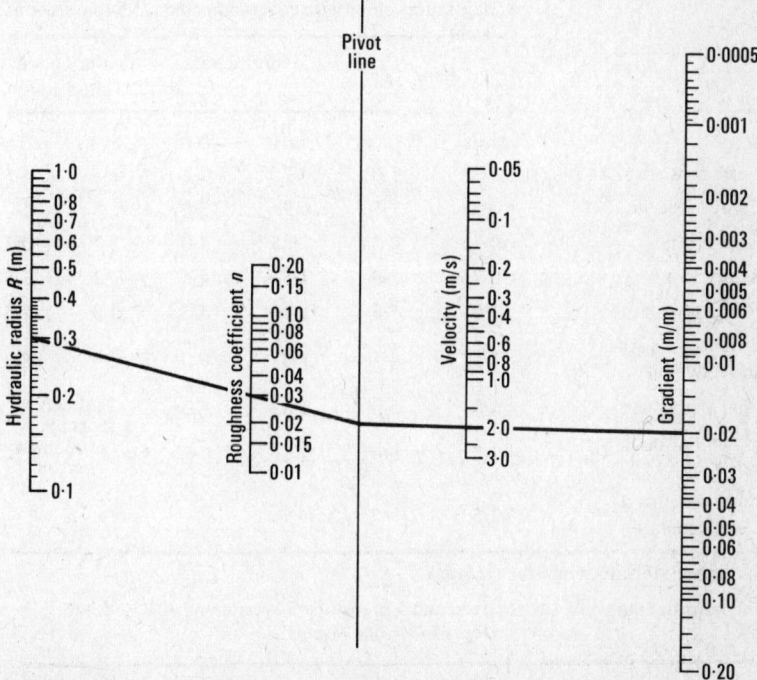

Figure 4.3. A nomograph for the solution of Manning's equation. If any three variables are known, the fourth can be found.
Example. Given $R = 0·3$ m, $n = 0·03$, gradient = 2 per cent or 0·02 m per m, find velocity V.
Solution. Join $R = 0·3$ and $n = 0·03$ and project to the pivot line. Join the point on the pivot line to gradient = 0·02. Intersection of the velocity scale gives $V = 2·0$ m/s.

fixed quantities, such as a given discharge to be carried, and some
variables which have restricted ranges, such as that the gradient should
be between certain limits or that the velocity must not exceed a given
value. Using this information the designer has to determine suitable
values for all the other variables. Very often there is no single unique
solution, but a range of possible solutions, and the designer is able to

Table 4.2

(a) Maximum non-scouring velocities for open channels

Material	Maximum velocity on cover expected after two seasons					
	Bare		Medium grass cover		Very good grass cover	
	m/s	ft/s	m/s	ft/s	m/s	ft/s
Very light silty sand	0·3	1·0	0·75	2·5	1·5	4·5
Light loose sand	0·5	1·5	0·9	3·0	1·5	5·0
Coarse sand	0·75	2·5	1·25	4·0	1·7	5·5
Sandy soil	0·75	2·5	1·5	4·5	2·0	6·5
Firm clay loam	1·0	3·5	1·7	5·5	2·3	7·5
Stiff clay or stiff gravelly soil	1·5	4·5	1·8	6·0	2·5	8·0
Coarse gravels	1·5	5·0	1·8	6·0	Unlikely to form very good grass cover	
Shale, hardpan, soft rock, etc.	1·8	6·0	2·1	7·0		
Hard cemented conglomerates	2·5	8·0	-	-	-	-

Intermediate values may be selected.

From Dept. of Conservation and Extension, Government of Rhodesia,
Dept. In-service Manual.

(b) Design velocities for grass waterways (m/s)

	Slope 0-5 per cent	5-10 per cent	10 per cent
Soil resistant to erosion	2·0	1·75	1·50
Erodible soils	1·75	1·50	1·25

select from a number of alternatives the combination of size of channel, gradient, velocity, etc. which is most suitable for the job in hand.

The general solution is to combine use of the continuity equation $Q = V \times A$ with use of Manning's equation, either to find V, or to select suitable combinations of hydraulic radius R, gradient S, and roughness n. It is not necessary to solve the equation mathematically, as graphical solutions such as Figure 4.3 have been prepared. Any of the four variables can be found if the other three are known.

For channels with earth sides, or lined with vegetation, a velocity of flow is required which is not so fast that it will scour the channel, nor so slow that sediment will be deposited. Some suitable velocities are given in Table 4.2 (a).

Real design problems are usually made much simpler by the fact that only a limited range of values is possible for each variable. For example, in designing unlined irrigation canals the gradient will be of

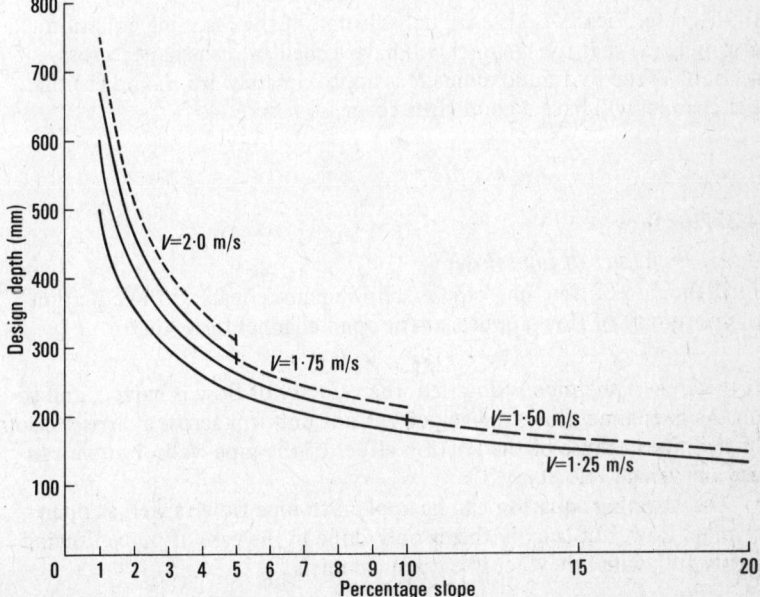

Figure 4.4.   Chart for obtaining the design depth of shallow grassed drains and waterways. Dotted lines show values which should only be used on soils resistant to erosion. Maximum permissable velocities are obtained from Table 4.2 (b).

the order of 1/1000, and there is no need to consider designs with gradients around 1/100.

Design of shallow grassed drains. A very common problem in soil conservation is the design of shallow grass-lined channels for draining surface run-off. In this case we know from experience that suitable designs will fall within certain limits, and so the following simplified design procedure may be used.

1. Estimate the maximum run off Q (m^3/s) using one of the methods in Section 3.1.
2. Choose the maximum permissible velocity V (m/s) from Table 4.2 (b).
3. For this velocity and the known ground slope find the channel depth d (m) from Figure 4.4.
4. Calculate the required channel width t (m) from

$$t = \frac{3 \times Q}{2 \times V \times d}$$

In this case, Figure 4.4 is a partial solution of the Manning equation assuming (a) that the channel will have a shallow dish-shaped cross-section, so the hydraulic radius R is approximately $1 \cdot 5\,d$, and (b) that the channel will have a good grass cover, so $n = 0 \cdot 04$.

4.3. Pipe flow

4.3.1. Principles of pipe flow

If the rate of flow in a pipe system remains constant, the equation of continuity of flow applies, as for open channel flow, that is

$$Q = A \times V.$$

If the area of the pipe is doubled, the velocity of flow is halved, and so on. As in channel flow, the velocity is not uniform across a cross-section of the pipe because of the friction effect of the pipe walls, but we can use an average velocity.

The Manning equation can be applied to pipe flow as well as open channel flow, but usually this is only done in the case of pipes flowing partly full, when they act like open channels.

4.3.2. Friction loss in pipes

When pipes are flowing full of water under pressure the relationship becomes more complicated because of the friction loss. The friction loss depends upon several factors.

The velocity of flow. The faster the flow the greater is the resistance. In fact the friction is proportional to the square of the velocity, so the resistance, which is small at low velocities, builds up quickly as the velocity increases.

The size of the pipe. In the Manning equation the hydraulic radius affects the velocity. In a full pipe of diameter D, the area $A = \frac{1}{4}\pi D^2$ and the wetted perimeter $P = \pi D$, so the hydraulic radius $R = A/P = \pi D^2/4\pi D = \frac{1}{4}D$, that is, the hydraulic radius is directly proportional to the pipe diameter. All other factors remaining the same, the bigger the pipe, the bigger the hydraulic radius and the bigger the velocity.

Roughness. A smooth pipe such as plastic has less frictional effect than a rough surface such as concrete. In some cases the roughness can change with age. Portable aluminium irrigation pipes get dented with use, and the resulting increase in roughness has to be allowed for when designing irrigation schemes (see Chapter 6). Galvanized iron and steel pipes may form rust or scale with age and this increases the roughness and reduces the velocity and the rate of flow.

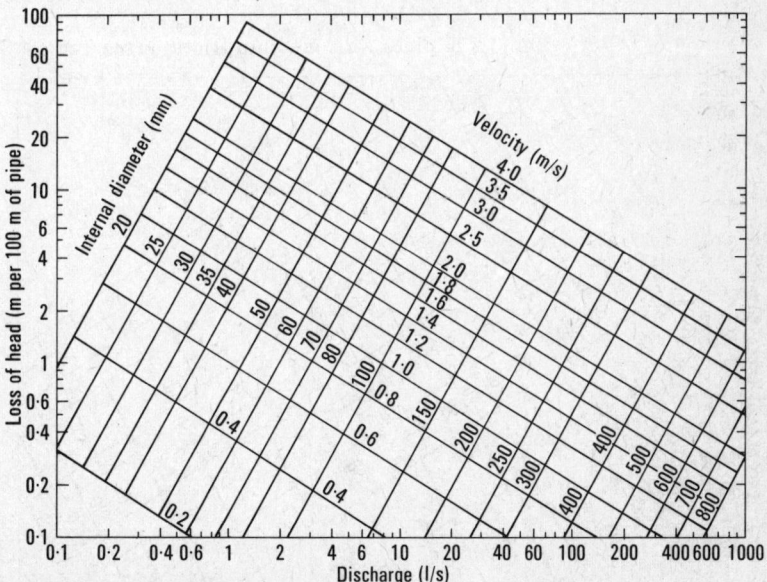

Figure 4.5.   Friction loss in pipes with low friction (see Table 4.3).

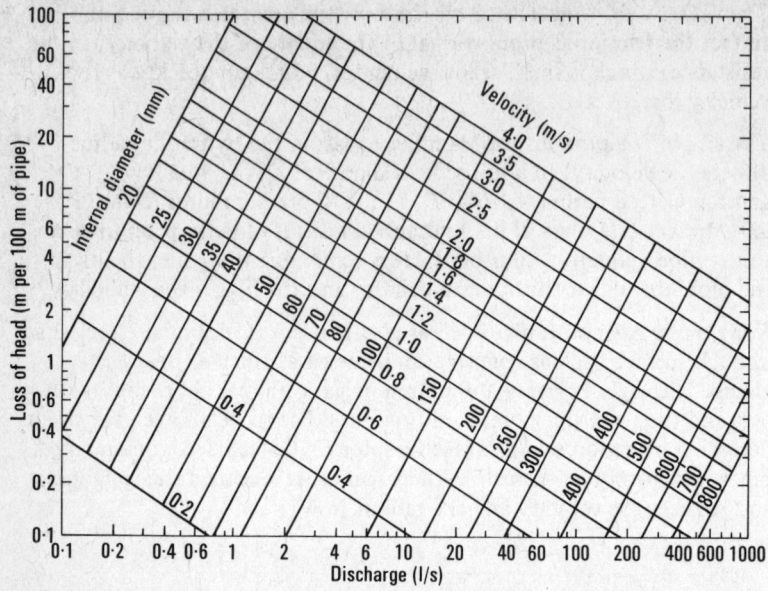

Figure 4.6.   Friction loss in pipes with medium friction (see Table 4.3).

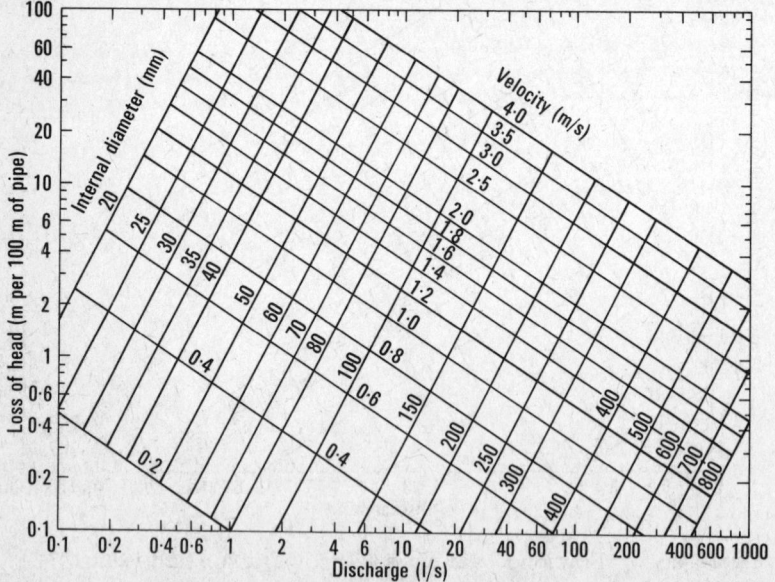

Figure 4.7.   Friction loss in pipes with high friction (see Table 4.3).

Table 4.3

Roughness characteristics of pipes of various materials

Roughness category	Material				
	Plastic	Asbestos/ cement	Aluminium	Galvanized	Concrete
Low friction (use Figure 4.5)	No constriction at joints, for example, taper-fit or solvent-welded	With smooth joints	Permanent	New	-
Medium friction (use Figure 4.6)	Some constriction at joints, for example, butt-welded	-	Portable	-	Spun concrete with gasket joints
High friction (use Figure 4.7)	-	-	-	Old	(a) Precast with gasket joints (b) Cast *in situ*

From F.A.O. Agricultural Development Paper 88.

Table 4.4

Approximate capacities of pipes of various sizes

Metric sizes (flow not exceeding 2 m/s)		English sizes (flow not exceeding 6 ft/s)	
Diameter (mm)	Flow (l/s)	Diameter (in)	Flow (gall/min)
50	3	2	48
75	8	2½	75
100	15	3	110
125	20	4	200
150	35	5	300
175	50	6	450
200	60	8	800
225	80	10	1250
250	100	12	1700
300	150	18	3700

Length. The frictional effect is directly proportional to the length of the pipe.

There are equations which relate these four variables, and allow the calculation of the friction loss for given conditions. Figures 4.5 - 4.7 give data for a range of pipe materials and sizes. The procedure is to determine from Table 4.3 which of the three classes of roughness is the one to use for a particular kind of pipe, then use the appropriate figure to read off the friction loss for a given pipe size and rate of flow. Table 4.4 shows the approximate capacity of pipes of various sizes.

4.3.3. Other losses

Whenever the flow of water in a pipe is interrupted or altered, such as by going round a sharp bend, or by going from one pipe size to another, there is a loss of pressure. The simplest explanation of this is that any such disturbance will cause turbulence in the flow, and this uses up energy, and so more energy has to be pushed into the pipe by using a higher pressure at the start of the pipe lines. This friction loss is proportional to the square of the velocity of flow, and while it can be ignored at low velocities such as those in drainage pipes, it can be significant in high-pressure irrigation schemes or water-supply systems, especially if large numbers of fittings are used. The effect can be expressed directly in terms of the pressure lost, but when the amount is small it is more convenient to express it as the length of straight pipe which would have the same friction loss. Table 4.5 gives some examples for irrigation and pipe fittings. An approximation often used in irrigation schemes is to add 10 per cent to the friction loss of the pipes to allow for all the miscellaneous minor losses.

Table 4.5
Head losses in pipe fittings

Pipe size (mm)	50 mm	75 mm	100 mm
	Length of straight pipe with similar head loss (m)		
Elbows and bends	1·25	1·75	3·0
T-junctions	3·0	4·5	6·0
Quick couplers Semi-quick couplers	1·50	1·5	1·5

4.4. Pumps

4.4.1. Pump principles

The design of pumping schemes and the selection of the pumps and motors is best left to engineers trained in these matters, but an explanation of the principles will be useful.

The two main factors affecting the selection of a pump are the amount Q to be pumped and the head or pressure H against which it must be lifted (the relationship between water pressure and head is explained in the Appendix). This head represents all the forces the pump has to over-come, and so it includes the suction lift to the pump, the change in elevation in the delivery pipe, the head loss due to friction in the pipes, any minor losses caused by pipe fittings, and finally the operating head, that is, the head equivalent to the pressure at which the water is dis-charged. Figure 4.8 shows how the total head is made up when pumping water for an irrigation scheme. The conversion of head of water to pressure, and the units, are explained in the Appendix.

For centrifugal pumps, the speed N of the pump also affects the discharge and the head. A useful way of defining the characteristics of a

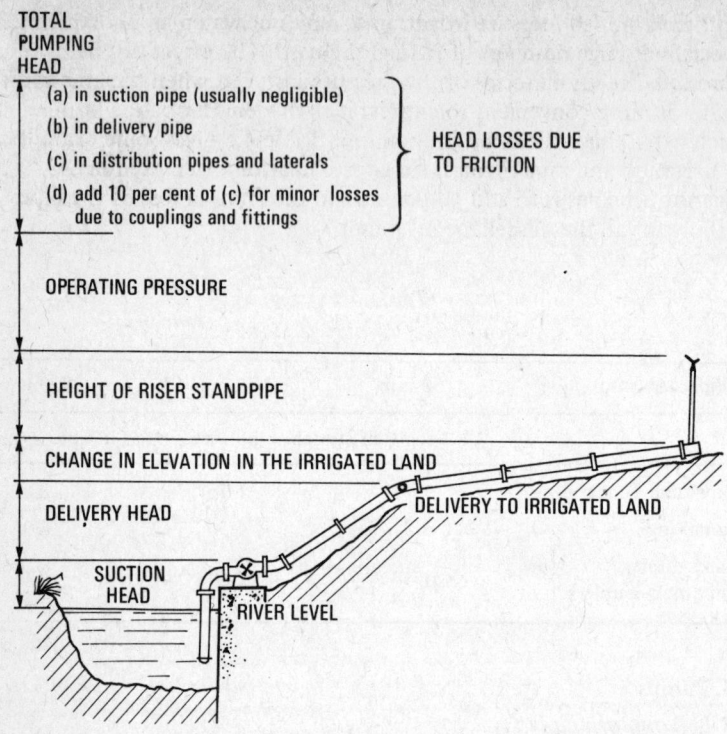

Figure 4.8.   In an irrigation scheme the total pumping head is made up of many components.

pump is by its *specific speed*, which is the speed at which the pump would run to deliver unit rate of flow (1 m³/hour or 1 gall/min) against a unit head (1 m or 1 ft). It is calculated from

$$\text{Specific speed } N_s = N \times \frac{Q^{1/2}}{H^{3/4}} \, ,$$

where N_s is specific speed in revs per minute (rpm), N is the speed in revs per minute, Q is discharge in cubic metres per hour (or gallons per minute in English units), and H is total head in metres (or feet in English units). Using these units, the British values will be the metric values divided by 1·273. The range of specific speeds for different kinds of pumps is shown in Figure 4.9.

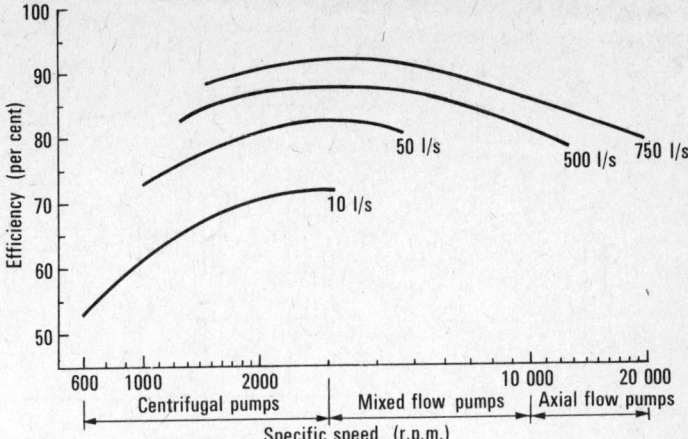

Figure 4.9.   The range of specific speed and efficiency for different kinds of pump.

The relationship between Q, H, and N are also given by what are sometimes called the affinity law:

$$Q \propto N,$$
$$H \propto N^2,$$
$$P \propto N^3.$$

This means that to increase the pump speed gives a directly corresponding increase in Q, but requires a big increase in the power required. For example, if the speed were doubled it would need 8 times more power, but the quantity would be doubled, and the head increased 4 times.

Because of these relationships the pump best suited for pumping large quantities against a low head is not the best for pumping small amounts against high heads. When a high head is required, more than one pump may be connected in series (Figure 4.10). Some pumps are in effect two or more pumps joined together in series, and these are called multi-stage pumps. For drainage work the requirement is usually large quantities and low lift, so several pumps are used in parallel (Figure 4.10).

The efficiency of a pump is the ratio of the work done by the pump in lifting the water compared with the work input from the engine or motor. The efficiency can vary from 45 per cent to 95 per cent (Figure 4.9). On the whole, pumps with a high specific speed are more

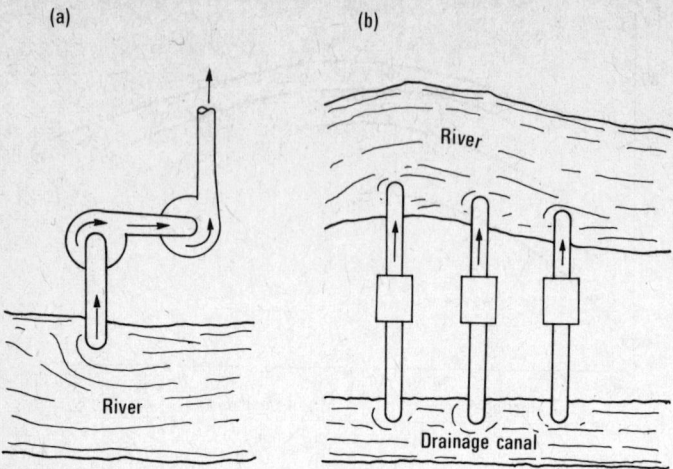

Figure 4.10.   Combining more than one pump: (a) in series for more lift; (b) in parallel for more volume.

efficient than pumps with a low specific speed, propeller pumps are more efficient than centrifugal pumps, and big pumps are more efficient than small pumps. This efficiency reflects only the losses in the pump itself, and to determine the efficiency of the whole pumping scheme one has to allow also for losses in the engine or motor, in transmission of power to the pump, and losses in transmission through the pipes.

Power. The power required at the pump depends on the discharge and the head, and can be calculated from the formula:

$$\text{Power required (kilowatts} = \text{kW)} = \frac{Q \times H}{100 \times E} \text{ ,}$$

Table 4.6
Loss of power in transmission

		Percentage
Direct-coupled electric motors add		10
Belt-driven electric motors	add	20
Direct-coupled I.C. engines	add	20
Belt-driven I.C. engines	add	25

where Q is the discharge in litres per second,
  H is the total head in metres, and
  E is the pump efficiency as a percentage.
In English units:

$$\text{Power required (horsepower} = \text{HP)} = \frac{Q\,(\text{gall/min}) \times H\,(\text{ft})}{33 \times E}$$

The loss of power in the transmission from the motor or engine to the pump is shown in Table 4.6.

4.4.2. Common kinds of pumps

The three kinds of pump used most frequently for drainage, irrigation, and water-supply schemes are *centrifugal pumps*, *propeller pumps*, and *mixed-flow pumps*, which combine the best features of the two others.

Centrifugal pumps. These are best for high-speed operation and pumping at high pressures. They can suck up water from 5 or 6 m below the pump, and are popular for sprinkler irrigation where a high pressure is required, and for water supplies when there is a high lift. An impeller rotating at high speed forces the water outwards and into the surrounding collecting chamber which leads to the discharge pipe. At the same time water is drawn in from the inlet to the centre of the impeller. There are two kinds of casing. The *volute* type (Figure 4.11 (a)) has a spiral casing with a cross-section which increases towards the outlet. The *turbine* type has stationary guide vanes round the impeller to give an improved flow pattern. The discharge is radial and at right angles to the inlet pipe (Figure 4.11 (b)).

(a)   (b)

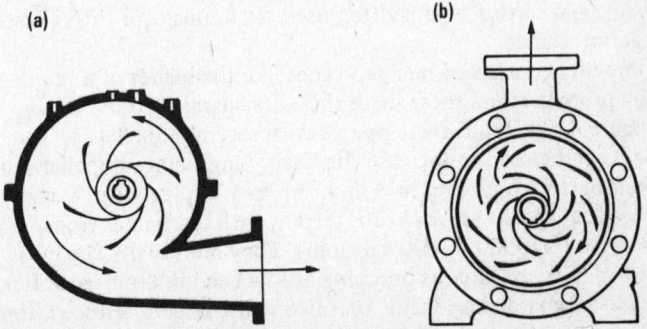

Figure 4.11.   Centrifugal pumps with (a) volute casing (b) turbine casing.

Small centrifugal pumps are usually mounted with the drive shaft entering the casing horizontally from one side, and the water inlet parallel on the opposite side of the casing. Larger pumps may be of the split-case double-suction type where the water is supplied to both sides of the impeller. There is a bearing for the drive shaft on either side of the impeller, so the thrust is balanced, and the symetrical design allows a more efficient construction. In multi-stage centrifugal pumps the discharge from the first impeller is led straight to the inlet of a second impeller mounted on the same drive shaft so that the discharge pressure is further increased.

A useful feature of centrifugal pumps is that, if the discharge is reduced, the pump and motor will not be overloaded. The pressure will rise to a fixed point which depends on the pump speed, and the system can be designed so that this pressure will not cause any damage.

The two disadvantages of centrifugal pumps arise on the suction or inlet side. Theoretically it is possible to draw water up the 10 m which corresponds to atmospheric pressure, but this would require perfect seals and in practice the maximum suction lift is about 6 m at sea level, with a reduction of 1 m per 1000 m of altitude. This limitation prevents the use of centrifugal pumps for pumping from deep wells and rivers with high banks.

The other disadvantage is that the pump has to be primed, that is, the whole of the inlet side has to be filled with water before the pump will work. Non-return valves are unreliable, so it is desirable to have some method for refilling the suction pipe if it has drained itself during a period of non-use.

Propeller pumps. These are best for pumping large quantities through small heads, and so this type is often used for drainage or lifting water into irrigation canals.

The impeller consists of inclined vanes like the blades of a fan or an aeroplane propeller, and these drive the water parallel to the drive shaft (Figure 4.12). Multi-stage pumps with several impellers on the same shaft can be used to increase the head. Single-stage impeller pumps are at their best with lifts up to 3 m, or up to 10 m for multi-stage pumps. Typical speeds are 450 - 1750 rpm, with discharges from 40 l/s to 4000 l/s (500 - 50000 gall/min). They are usually the most efficient type for continuous pumping at low head in drainage schemes. A danger to be guarded against is that if the discharge is reduced, the power load increases, and an electric motor could be burned out by the accidental closing of a discharge valve.

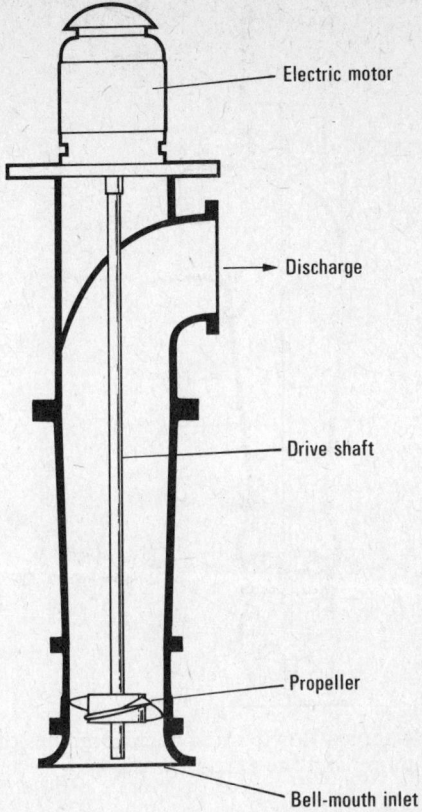

Figure 4.12.   An axial-flow or propeller pump.

Mixed-flow pumps. These are a cross between the centrifugal and propeller types in order to combine the best of each (Figure 4.13). Deep-well turbine pumps for pumping from deep wells are usually of this type. Several stages are possible on the same shaft. Vanes are fitted in the casing to straighten out the flow from each stage, hence the description as turbine pumps. The only limitation to the depth of pumping from deep wells is the length of the drive shaft. The dis-advantage is that maintenance and inspection of the impellers and bearings is difficult.

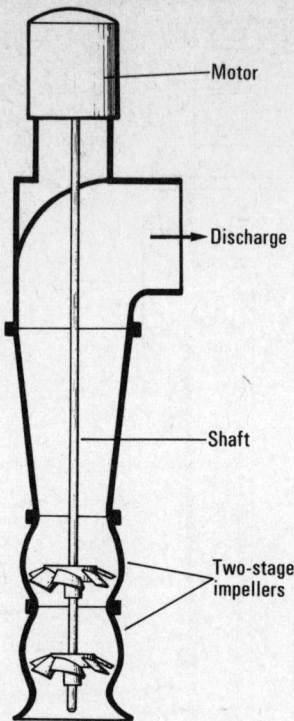

Figure 4.13.   A mixed-flow pump which combines the advantages of the centrifugal pump and the axial-flow pump.

Less common pumps. Some other less commonly used pumps are the following.

1. *Reciprocating pumps.* These can generate very high pressures, as in the case of injector pumps for diesel engines, but they have a low rate of discharge and high cost.

2. *Diaphragm pumps.* The liquid is pumped by the oscillation of a flexible diaphragm which acts rather like the piston of a reciprocating pump. A common example is the mechanical fuel pump on a petrol-engined car. At the other end of the size range, large diaphragm pumps with a diaphragm up to 0·5 m in diameter are used for de-watering excavations because of their special advantage of being able to pump muddy water (Figure 4.14).

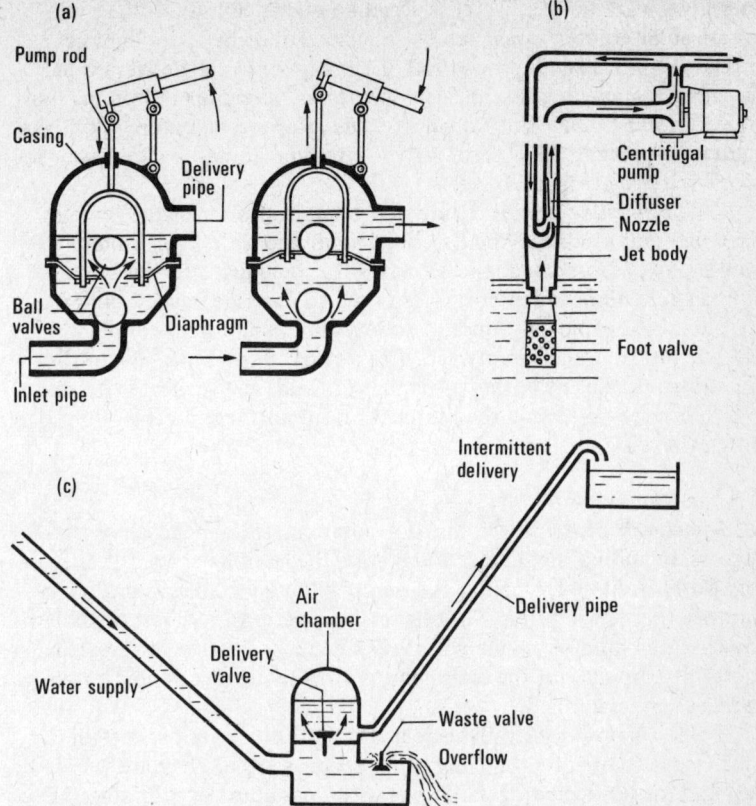

Figure 4.14.   Some special-purpose pumps: (a) diaphragm pump (From Culpin, *Farm machinery*, Crosby Lockwood and Son, London.) (b) jet pump; (c) hydraulic ram.

3. *Jet pumps.* Water is pumped down a pipe in a well and through an upward-facing jet so that an increased flow is drawn up the return pipe. Some of this flow is pumped back down to the jet, and the rest goes into the delivery system. The advantage is that there are no moving parts down the well, but only small discharges are possible. It is useful for rural water supplies (Figure 4.14).

4. *Hydraulic rams.* No external source of power is required for this kind of pump which instead relies on the energy of a permanent stream

in a steep watercourse. Water is piped down the hillside to the pump, where an alternately open and closing valve uses the water hammer effect to 'ram' a small proportion of the flow into the delivery pipe. Because of shock loading and high pressures the pump and piping must be very robust. Only small quantities of water can be pumped, but the hydraulic ram can be a useful way of pumping domestic supplies or for stock-watering (Figure 4.14).

   5. *Submersible pumps.* There is some confusion of names here as two different kinds may be described as submersible. Any kind of deep well pump is submerged, and so may with some justification be called a submersible pump. The term is also used to describe what should really be 'submersible-motor pumps', where a pump and direct-coupled electric motor are all submerged. The difficulty lies in water-proofing the motor, while the advantage is that it avoids the vertical drive shaft which is necessary when the motor is on the surface and the pump is down the well.

4.4.3. Engines and motors

   The choice of the power unit for pumping will depend on several factors, including the initial capital cost, the running costs, the speed and power required, whether the pump will be portable or stationary, and whether it will pump for long or short periods. It is not possible to say that one kind of engine will always be most suitable for a particular kind of pumping, but the main features of the most common power sources can be defined.

   If electric power is available, electric motors are almost certain to be cheaper to operate than any other power source. They are very suitable for permanent installations with long hours of pumping. They are not suitable for portable pump units, nor for pumping for short intermittent periods. The best arrangement is when the motor is coupled directly to a high-speed pump. Belt drives are undesirable as they dissipate the important advantage of a simple trouble-free installation. Most electric motors work at one of the standard speeds, which are shown together with the power requirements in Table 4.7. The power rating quoted by manufacturers is the safe output for continuous operation after allowing a 10 or 15 per cent safety factor.

   Petrol engines have a low initial cost but high running costs. They have relatively short life, and high maintenance requirement. They operate equally well over a wide speed range. They are therefore most suitable for portable units, and for intermittent pumping for short periods, and as a stand-by unit for use when the main pumps are being

Table 4.7
Electric motors: speeds and power requirements

Electricity supply	Common speeds (rpm)			
50 cycles per second (Hz)	725	960	1450†	2900
60 cycles per second (Hz)	870	1160	1750†	

Power output of motor	Electricity supply required	
Up to 5 kW (7 HP)	110 V or 220 V	Single phase
5 - 20 kW (7 - 25 HP)	220 V	Three-phase
More than 20 kW (25 HP)	440 V	Three-phase

† Most common speeds.

overhauled. Petrol engines are usually used only up to 20 or 30 kW (30 - 40 HP), beyond which diesel engines are better.

Diesel engines have a higher initial cost but lower running cost. They are usually rugged in construction and built for long trouble-free service in tough conditions. Large engines are available, but above about 75 kW it may be preferable to have more than one smaller engine than a single very large one. The most economical operation is at constant speed, and usually at a lower speed than petrol engines. Diesels are heavier per kilowatt than petrol engines, and so less suitable for portable units, although the design of small light engines has progressed a great deal in recent years.

The rated power quoted by manufacturers may be either for continuous output in operating conditions, or the bench-test output of a stripped engine without fan, pump, air cleaner, etc. The difference can be up to 25 per cent, so it is important to check which kind of rating is quoted. Internal combustion engines lose power at high altitudes and in high temperatures. The loss of power is 1 per cent per 100 m above 100 m and 1 per cent per 5 °C above 15 °C. For economical long life, petrol and diesel engines should not be operated continuously at their maximum power output, but at 70 - 80 per cent of this figure. For all of these reasons, when choosing an engine it is wise to have much more spare power in hand than in the case of an electric motor.

Most tractors have a *PTO (power take-off)* from which power can be taken to a pump through a flat belt or V-belt or drive shaft. About 60 per cent of the rated power of a tractor is available at the PTO. Centrifugal pumps are available which can be mounted on the tractor, and this gives a completely self-contained mobile pumping unit. This can be very useful for short-period pumping, for example, in irrigation. If the pumping is from a river or canal, the ability to move the engine and pump easily to any point on the bank may allow a saving in the amount of piping required. If intermittent pumping is all that is required, the tractor can then be used elsewhere when not pumping, so it may avoid buying an engine for very limited use. However, a self-contained self-propelled pump unit is only economically justified when its mobility will be useful. If the requirement is for a lot of pumping in one place, a stationary engine will be cheaper than a tractor.

5. Water storage

5.1. Earth dams

We have already said that water is the key to agricultural development in most countries. Water storage is important because the water that is available is often not there at the best time. The way to even out the floods and droughts is to store the flood water while there is more than enough and use it later when there is less than enough.

For some purposes, such as flood control, irrigation, or power generation, water can be stored most cheaply in large dams which take advantage of favourable sites. There are other purposes for which storage in small dams and reservoirs may be better. They can be sited nearer to where the water is required, and they can be built with less money, less equipment, and less engineering skill. They can be used as an instrument of government policy, for example, by construction at subsidized cost to help a developing region.

The design and construction of dams and weirs does call for engineering skills, but it is unrealistic to expect that a qualified engineer can be available for every job, no matter how simple. A system of graded responsibility is the answer, such as an example from Africa in which:

(1)   dams up to 3 m in height may be built by field advisory officers using standard designs procedures such as those in this chapter;

(2)   between 3 m and 5 m the advice of a senior experienced field officer must be obtained; and

(3)   over 5 m an engineer must be called in.

5.1.1. Planning earth dams

The first step in planning a dam is to define quite clearly the purpose of the dam, so that design requirements can be identified. For example, if a cattle watering point is required in order to make use of grazing in a particular area this means there is no point in looking for a site outside the defined area. If the water has to be available throughout the year this means there must be a minimum depth so that water is always available after allowing for evaporation, but on the other hand the amount of water required for stock-watering will be much less than for an irrigation storage dam. When the purpose is irrigation this means the amount of storage is defined, and also the period when the water must be available. When the purpose is clear, the consideration of the other following factors will be easy.

Site location. This is usually dictated by the purpose, as in the above example of providing water in a particular grazing area. When the purpose is irrigation the location will be dictated by the answers to questions such as:

'Can the scheme be anywhere in the district, or is there a particular piece of land to be irrigated and which the dam has to be near?'

'Is it a gravity scheme where the dam must be above the land, or can we pump from below?'

'Could we have the dam site upstream and bring the water down in the stream?'

'If the dam is at some distance from the scheme would the supply canal have to be lined?'

Capacity. In order to plan the capacity of the dam, we need to ask 'How much water will be used from the dam, and how much will the demand fluctuate through the year? What is the maximum demand? What are the consequences if the supply dries up? Is a carry over from one year to the next necessary or desirable? What will the losses be from evaporation and from seepage?'

Site selection. The requirements of a good site must be considered under the headings of storage, spillways, and soil.

1. *Storage basin.* The shape of the valley influences how much will be stored for a given height of dam wall. The ideal site has a long throwback, that is the surface of the stored water goes a long way back upstream giving a big capacity. It will also have steep valley sides at the dam site so the earth wall is short. A valley with gently sloping sides means a long and expensive wall. If evaporation is high, it is better to have deep water and a small surface area rather than a large surface area of shallow water. Sites with good basin storage are often found where the stream passes from a wide valley through a narrow gorge, or where the streambed changes from a flat to a steeper gradient, or at the confluence of two streams so that there are two storage basins for one wall.

2. *Spillways.* It is unusual for a dam site to be able to store the whole of the run-off, and the spillway is the channel or pipe provided to allow the surplus to pass when the dam is full. The cost of small earth dams is greatly increased if spillways have to be built of concrete or similar materials, so a major factor in site selection is looking for

sites where the overflow can be safely discharged over grass-covered spillways. The size of spillway depends on the maximum expected flood and this in turn depends on the size of the catchment area and the rainfall, so it is wise to limit the maximum size of catchments for dams with grassed spillways to

250 ha or 1 mile^2 where mean annual rainfall is more than
625 mm or 25 in,

500 ha or 2 mile^2 where mean annual rainfall is less than
625 mm or 25 in.

3. *Soils.* The site must have suitable soil conditions for constructing an earth dam. The soil of the basin should not be porous, and the site for the wall should be free from boulders and termite mounds. Suitable soil for constructing the wall should be available nearby. The best kinds of soil for different parts of the wall are discussed in Section 5.1.5 under *Construction sequence.* Certain soils are unsuitable for the construction of dams; these include

saline, alkaline, or sodic soils, or any soils with abnormal chemistry,
peat or other soils high in organic matter,
heavy clays subject to swelling, shrinking, and cracking,
light sands,
soils containing a high proportion of fine silt.

5.1.2. Estimating quantities

In order to choose between alternative sites, or between alternative sizes of dam, it is necessary to make estimates of the storage capacity, the earthworks, and the cost. A common practice is to make a quick estimate in order to eliminate the least attractive possibilities, and then to use a more accurate method on those which show most promise.

Storage capacity

1. An approximate estimate of the capacity is

$$Q = \frac{L \times T \times D}{6} \, ,$$

where Q is the capacity in cubic metres,
L is the length of the dam wall in metres at full supply level,
D is the maximum depth in metres, and
T is the throwback in metres.

This assumes that the basin is a pyramid whose base is the dam wall (Figure 5.1 (a)).

In English units

$$Q = L \times T \times D,$$

where L, T, and D are in feet, and Q in gallons.

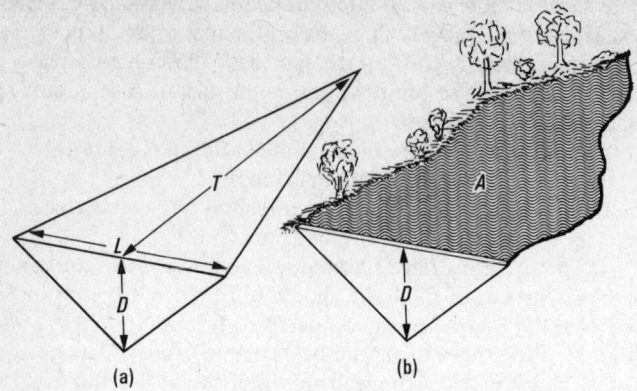

Figure 5.1.   Estimating the storage capacity of a reservoir.

2. A slightly more accurate estimate is obtained by measuring the surface area at full supply level, and using the formula

$$Q = \frac{A \times D}{3},$$

where D is the maximum depth in metres and
 A is the surface area in square metres.
This assumes that the basin is a pyramid whose base is the water surface (Figure 5.1 (b)).
In English units

$$Q = 2 \times A \times D,$$

where A is in square feet, D in feet, and Q in gallons.

3. To obtain more accurate figures it is necessary to make a contour survey of the basin. The lake boundary is drawn on the map for various depths of water at say half metre intervals, up to full supply level. For each layer, the volume is calculated from the depth of the layer times the average of the upper and lower area.

Quantity of earthworks
1. The approximate formula is

$$V = 0 \cdot 216 \, HL(2C + HS),$$

where V is the volume in cubic metres,
 H is the maximum height of wall in metres,
 L is the length of wall in metres along the crest,
 C is the crest width in metres, and
 S is the sum of the upstream and downstream slopes of the

embankment, for example, if the upstream slope is 3 to 1 and the downstream slope is 2 to 1, $S = 5$.

In English units
$$V = \frac{8HL(2C + HS)}{1000} ,$$

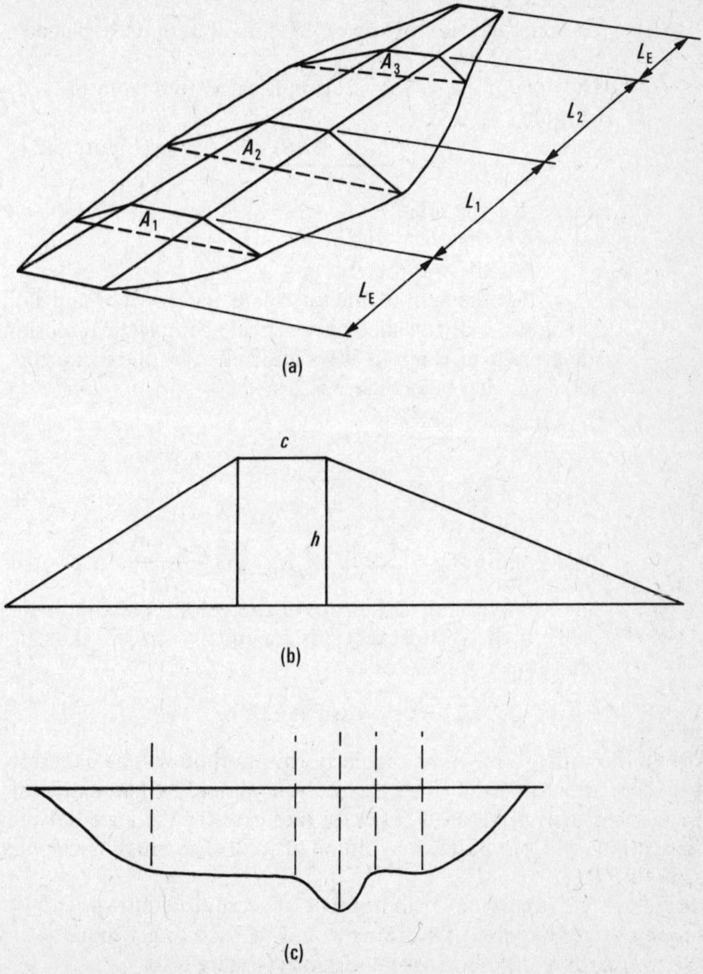

(a)

(b)

(c)

Figure 5.2. Estimating the volume of an earth dam by the method of vertical slices.

where V is in cubic feet, and H, L, and C in feet.

More accurate estimates are obtained by considering the embankment as a series of slices and calculating the volume of each piece.

2. Calculation between vertical sections (Figure 5.2):

(a)   The cross-section of the valley along the centre line of the dam is plotted.

(b)   The cross-section of the embankment at equally placed points is plotted.

(c)   The area of each cross-section is calculated from the formula

$$A = \frac{h(2c + hS)}{2} , \qquad \text{(Figure 5.2 (b)),}$$

where A is the area

c is the crest width,

h is the wall height, and

S is the sum of the upstream and downstream slopes.

(d)   The volume of the slice between adjacent vertical sections is the average of the two areas multiplied by the distance between, that is in Figure 5.2 (a),

$$V_1 = \frac{A_1 + A_2}{2} \times L_1$$

and $V_2 = \dfrac{A_2 + A_3}{2} \times L_2$.

At the ends $V_E = \dfrac{A_1 + 0}{2} \times L_E$ and $\dfrac{A_3 + 0}{2} \times L_E$.

(e)   The total volume is the sum of the volumes of the vertical slices. If the vertical sections are equally spaced, the calculation simplifies to

$$V = L (A_1 + A_2 + A_3 + \dots) .$$

When the valley sides slope regularly this method will be accurate, but if there are changes in slope the sections should be taken at the point of change as in Figure 5.2 (c). In that case the distance between cross-sections will vary, and the volume of each slice must be calculated separately.

In Figure 5.2 (a) only a small number of sections is shown, and in practice a larger number of sections would be used to get more accuracy.

3. Calculation from horizontal sections (Figure 5.3).

(a)   The cross-section of the valley along the centre line of the dam is plotted (Figure 5.3 (a)).

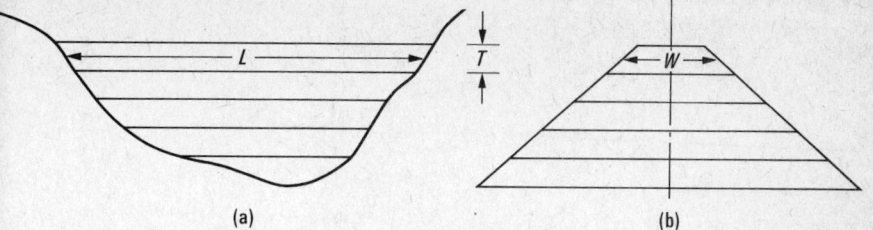

Figure 5.3.   Estimating the volume of an earth dam by the method of horizontal slices.

(b)   The cross-section of the wall at the deepest point of the valley is plotted alongside using the same vertical scale but an enlarged horizontal scale (Figure 5.3 (b)).

(c)   The wall is divided into horizontal sections of equal thickness T. The length L at the mid-point of the section is measured on Figure 5.3 (a) and the width W at the mid-point of the section is measured on Figure 5.3 (b) (remembering that the horizontal scales of the two diagrams are not the same).

(d)   The volume of each section is obtained from
$$V = W \times L \times T.$$

(e)   The total volume is the sum of the volumes of the horizontal slices.

As with the previous method the accuracy is increased if a large number of thinner slices is used for the calculation.

5.1.3. Spillways

Few dams on streams and rivers can be built big enough to store all the run-off, and some provision is required to pass on the surplus flood water after the dam has filled. The spillway is designed to do this so that the floods cannot cause damage to the dam. The problem is that storing the water in the reservoir raises the level from which the water is discharged, thus giving it additional energy which has to be dispersed without causing erosion. There are several different kinds of spillway.

Cut spillways. For small conservation dams one solution is open channel spillways cut into the bank at the side of the dam wall (Figure 5.4). These are cheap and effective, but there is a risk of erosion if the water flows too quickly. If it is possible to maintain a good grass cover, this greatly reduces the required size of spillway, but grassed channels

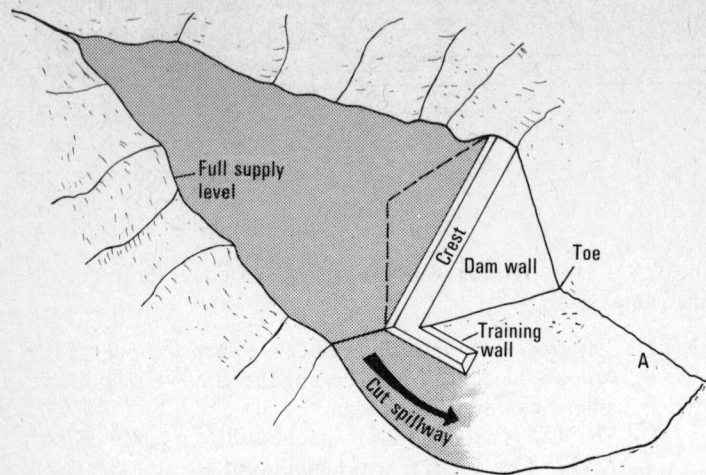

Figure 5.4.   Plan of a small earth dam with a cut spillway.

do require careful design, construction, and maintenance, as discussed in Chapter 4.

The most vulnerable point on cut spillways is where the water re-joins the stream. To prevent erosion at this point (A in Figure 5.4) the stream bank must be cut back to a gentle slope and planted with grass like the rest of the spillway. Alternatively a concrete or masonry drop structure may be provided, as discussed in Section 8.4.2.

It is undesirable for grass spillways to be constantly saturated. This makes them more vulnerable to erosion because the types of grass which are most suitable for lining channels do not like continually wet conditions. So if a stream is likely to have a small constant flow, or one which continues a considerable time after each storm, a *trickle flow outlet* should be provided. This can take the form of a small brick-lined or concrete channel set in the spillway, or it can be a small diameter pipe going through the dam wall, preferably on the opposite side of the stream from the spillway.

If it is not practical to protect a cut spillway with grass, it may be necessary to protect it, or perhaps the most vulnerable parts, with stone pitching or concrete.

Natural spillways. These are used where the site conditions make it possible to divert the flood flow into a naturally existing waterway or channel. Figure 5.5 shows an example where the flood is diverted over a

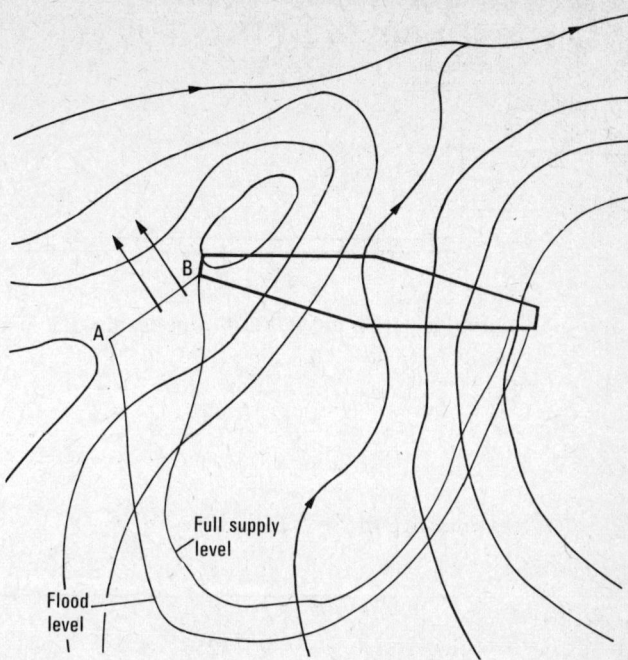

Figure 5.5.   A reservoir with a natural spillway into an adjacent water-course.

crest into the adjacent valley. The crest should have a gentle slope between A and B so that the depth of flow is small, and the slope down to the stream should be gentle and have enough grass cover so there is no risk of erosion.

Mechanical spillways. These use pipes to pass the flood through the embankment. It would not be practical to have the pipe large enough to pass the maximum peak flow, so temporary flood storage is provided above the pipe outlet. This fills up during the height of the flood, and discharges through the pipe after the peak flow has passed. (Figure 5.6). An emergency spillway, usually a grassed cut spillway, is also provided for the exceptional floods.

Drop inlets. These are often used at the entry to pipe spillways (Figure 5.6), partly to ensure a good discharge through the pipe and partly so

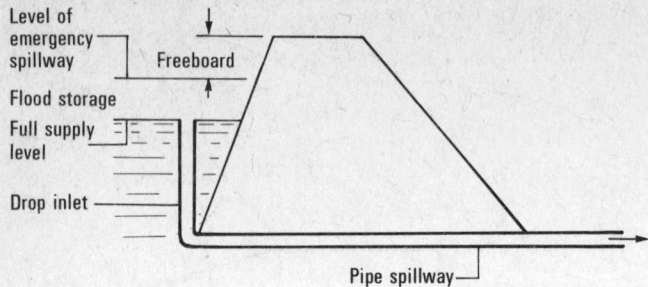

Figure 5.6.   A reservoir with a mechanical (pipe) spillway and an emergency spillway.

Table 5.1
Flood discharge factor F for spillway design

(a) Metric units

Freeboard (mm)	Area (ha)							
	50	100	150	200	250	300	400	500
750	28	56	85	113	140	170	225	280
1000	18	35	53	70	88	105	140	175
1250	13	25	38	50	63	75	100	125
1500	10	20	30	40	50	60	80	100

(b) English units

Freeboard (ft)	Area (acres)									
	100	200	300	400	500	600	800	1000	1200	1400
3·0	55	110	165	220	275	330	440	550	660	770
3·5	43	85	128	170	212	255	340	425	510	595
4·0	35	70	105	140	175	210	280	350	420	490
4·5	30	60	88	118	145	175	230	290	350	405
5·0	25	50	75	100	125	150	200	250	300	350

From Dept. of Conservation and Extension, Government of Rhodesia, *Handbook of basic instructions for dam construction.*

that the pipe can be buried below the dam wall, as discussed in Section 5.1.5.

Design of spillways. The maximum flood is quite likely to occur when the dam is already full, so provision must be made either to pass the whole flow over the spillway, or to cushion the flood with temporary storage and a reduced flow for a longer period. In tropical and sub-tropical countries the large volume of flash floods from isolated storms makes it costly to use the temporary storage method, and so for small cheap conservation dams the grass spillway which can pass the maximum flood is usually best.

The following design procedure has been widely used in southern and central Africa in conditions when a good grass cover can be

Table 5.2
Rainfall intensity factor R for spillway design

(a) *Metric units*

Length of catchment (km)	Mean annual rainfall (mm)		
	400	800	1200
1	0·83	0·90	0·93
2	0·62	0·66	0·68
3	0·49	0·52	0·54
4	0·40	0·43	0·45
5	0·34	0·37	0·38
6	0·29	0·32	0·33

(b) *English units*

Length of catchment (miles)	Mean annual rainfall (in)		
	16	32	48
0·5	0·90	0·96	1·00
1	0·69	0·73	0·75
1·5	0·56	0·59	0·60
2	0·47	0·50	0·51
2·5	0·40	0·43	0·44
3	0·35	0·38	0·39
4	0·28	0·30	0·31

established. The method is closely related to Cook's method of estimating maximum run-off (Chapter 3), but leads directly to the required spillway size.

1. The *flood discharge factor F* is found from Table 5.1 for the catchment area and the proposed freeboard, that is, the height of dam wall above the spillway.
2. The *rainfall intensity factor R* is found from Table 5.2 using the length of catchment and the mean annual rainfall.
3. The *topographical factor T* is found from Table 5.3 (this is the same as finding the catchment characteristics for Cook's method).
4. The required width *W* (in metres) of spillway is
$$W = F \times R \times T.$$

A range of alternative designs can be obtained. Increasing the design freeboard in Table 5.1 will give a reduced width, or reducing the freeboard will increase the required width.

The detailed design of dams with mechanical spillways and temporary storage is more complicated, and uses the technique of flood

Table 5.3

Topographical factor T for spillway design

Surface cover = a	Soil type = b	Slope = c
(i) Thick bush 0·05	(i) Deep, well-drained soils 0·10	(i) Very flat to gentle 0·05
(ii) Heavy grass 0·10	(ii) Deep, moderately pervious soil 0·20	(ii) Moderate 0·10
(iii) Scrub or medium grass 0·15	(iii) Soils of fair permeability and depth 0·25	(iii) Rolling 0·15
(iv) Cultivated lands 0·20	(iv) Shallow soils with impeded drainage 0·30	(iv) Hilly or steep 0·20
(v) Bare or eroded 0·25	(v) Medium-heavy clays or rocky surfaces 0·40	(v) Mountainous 0·25
	(vi) Impervious surfaces and waterlogged soils 0·50	
	$T = a + b + c$	

From Dept. of Conservation and Extension, Government of Rhodesia, *Handbook of basic instructions for dam construction.*

Table 5.4
Capacities of mechanical pipe spillways

(a) Metric units

Diameter of Spillway pipe (mm)	Approximate Capacity (l/s)	Diameter of Drop-inlet pipe (mm)
200	5600	200
250	7100	250
300	8500	300
400	11300	500
500	14200	600 - 750
600	17100	750 - 1000

(b) English units

Diameter of Spillway pipe (in)	Approximate Capacity (ft^3/s)	Diameter of Drop-inlet pipe (in)
8	2·5	8
10	5·0	10
12	7·5	12
15	15·0	18
18	20·0	21 - 24
21	30·0	24 - 36
24	40·0	30 - 42

From Schwab, Frevert, Edminster, and Barnes,
Soil and water conservation engineering, Wiley, New York.

routing which can be found in most text-books on engineering
hydrology. Basically the method is to draw up a continually changing
balance-sheet between the incoming flood, the outflow from the pipe
spillway, and the available storage. A suitable combination of pipe size
and storage can then be chosen. Table 5.4 gives some pipe sizes and
capacities. To make sure that the pipe flows full, the cross-sectional
area of the drop inlet is always bigger than that of the pipe.

5.1.4. The embankment or dam wall

Impermeable core. Seepage through or under the dam wall must be prevented, or reduced to the point where it is not important. Unless the whole wall can be built of impervious material, a central impermeable core is required as shown in Figure 5.7. The core consists of clay, compacted and consolidated as described in Section 5.1.5. The core should be keyed into the foundation so there is no possibility of seepage below it, and it must be continued above full supply level along the length of the wall and at both ends.

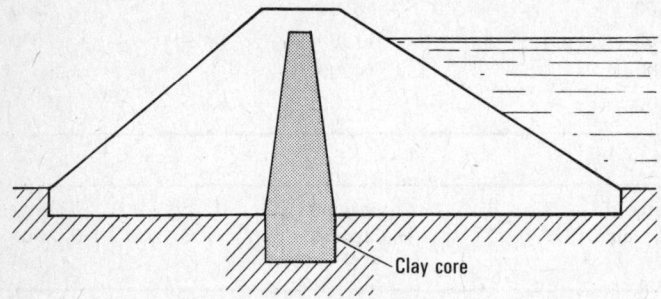

Figure 5.7.   An earth dam with an impermeable clay core.

Side slopes. The embankment walls must slope inwards at an angle which will be stable at all times. Saturated soil is less stable than dry soil, so the water side of the wall has to be flatter than the downstream side. For most soils the slopes should be 3:1 on the upstream side and 2:1 on the downstream. For soils approaching the ideal mixture of grain sizes and which are carefully placed and well compacted, the slopes could be increased to 2:1 upstream and 1½:1 downstream. Soils which will be difficult to compact should have slopes of 3:1 or 4:1. For small dams the slope does not make much difference to the total earth moving costs and a conservative slope will not increase the cost much. For large dams, if it is necessary to use flatter slopes because of less suitable soil, this will make a big difference to the volume as shown in Figure 5.8 (a).

Crest width. The top of the dam wall must be wide enough for both the construction equipment and for any traffic if the wall is going to be used as a bridge. Unnecessary width will add to the volume of earthwork required as shown in Figure 5.8 (b). If the wall will only carry foot

(a)

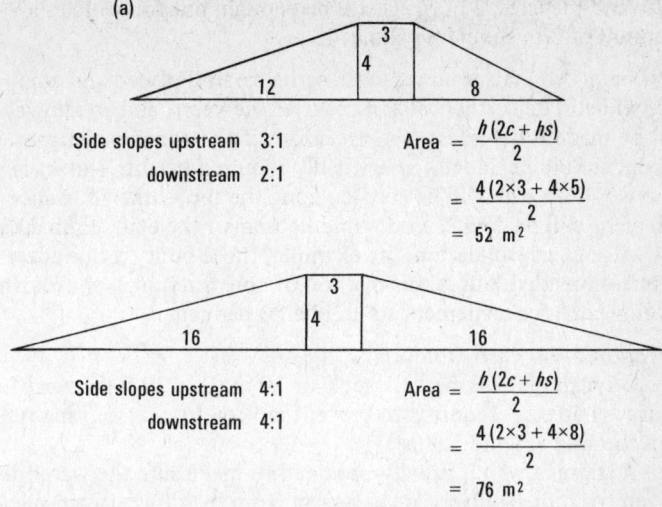

Side slopes upstream   3:1

downstream   2:1

$$\text{Area} = \frac{h(2c + hs)}{2}$$

$$= \frac{4(2 \times 3 + 4 \times 5)}{2}$$

$$= 52 \text{ m}^2$$

Side slopes upstream   4:1

downstream   4:1

$$\text{Area} = \frac{h(2c + hs)}{2}$$

$$= \frac{4(2 \times 3 + 4 \times 8)}{2}$$

$$= 76 \text{ m}^2$$

(b)

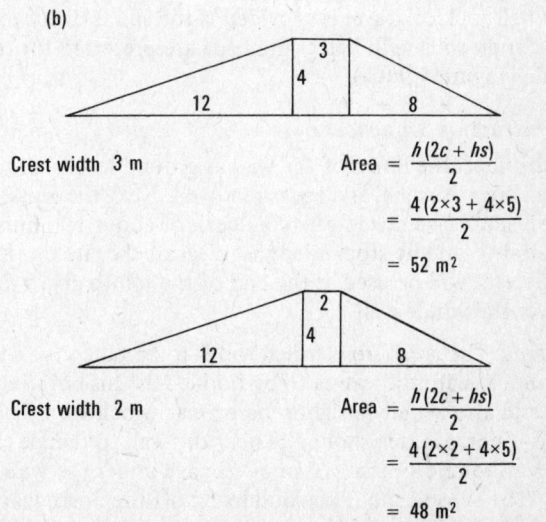

Crest width 3 m

$$\text{Area} = \frac{h(2c + hs)}{2}$$

$$= \frac{4(2 \times 3 + 4 \times 5)}{2}$$

$$= 52 \text{ m}^2$$

Crest width 2 m

$$\text{Area} = \frac{h(2c + hs)}{2}$$

$$= \frac{4(2 \times 2 + 4 \times 5)}{2}$$

$$= 48 \text{ m}^2$$

Figure 5.8.   The effect of side slopes and of crest width on the volume of an earth dam.

traffic and bicycles, 2 m (6 ft) will be enough, but for vehicles a minimum of 3 m should be allowed.

Settlement. No matter how carefully the earth is placed and compacted, there will still be further settlement over the years, and an allowance must be made for this. For average soils with reasonable compaction 10 per cent should be added to the finally required height. This means that the newly built wall will be convex along the top as the allowance for settlement will be greater in the middle than at the ends. Embankments built without consolidation, for example, those built by bulldozers, are not recommended, but if this method of construction is unavoidable the allowance for settlement should be 20 per cent.

Protection of downstream wall.

1. On high embankments, over 8 or 10 m (25 - 30 ft), a *berm* or drainage channel is required to prevent erosion by intercepting run-off down the wall (Figure 5.9 (a)).

2. A training wall is usually provided to make sure the water discharged from the spillway is kept away from the downstream side of the wall (Figure 5.9 (b)).

3. To keep the downstream toe of the wall properly drained a toe drain of small rock or gravel is provided if the soil is likely to have poor drainage. Sandy soils with better drainage are preferred for the downstream side (Figure 5.9 (c)).

5.1.5. Construction sequence

Site preparation. The limits of the wall at ground level should be staked out, and all trees, shrubs, and roots removed. Next the grass, grass roots, and topsoil should be dozed off to a depth of about 100 mm (4 in). This material should be stockpiled just clear of the site on the downstream side as it will be used at the end of the job to give a dressing of topsoil over the whole wall.

Borrow areas. The areas from which soil is to be taken for the embankment should be within the area to be flooded as this both increases the capacity, and also avoids unsightly bare areas which would be liable to erosion. The borrow area should be near the wall to reduce the haul distance, but not closer than 10 m or seepage under the wall may be increased. The topsoil should be bulldozed off the borrow areas, and stockpiled like the topsoil from the dam-wall site.

Outlet pipes. It is always difficult to consolidate earth which has to be placed around a pipe already in position. To get good compaction means

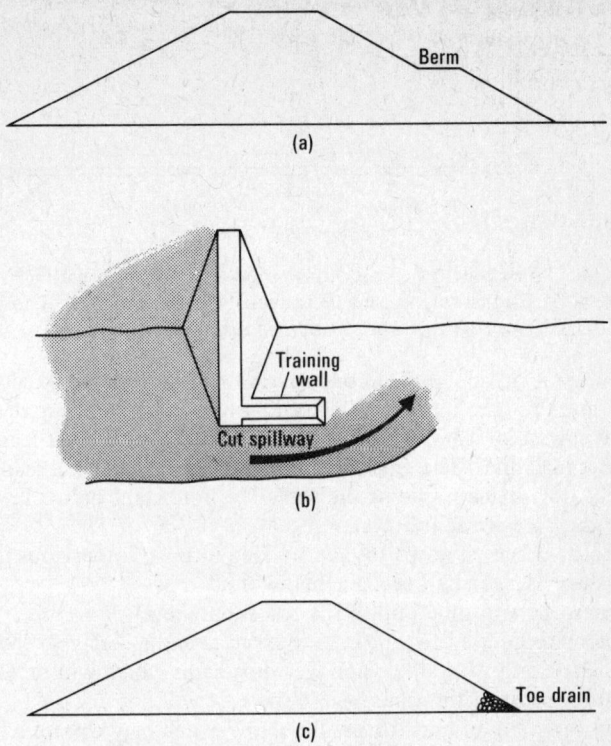

Figure 5.9.   Protecting an earth dam wall: (a) a berm on the down-stream face; (b) a training wall; (c) a toe drain.

using methods which are liable to damage or dislodge the pipe. The better method is to put the pipe in undisturbed soil under the dam wall, as shown in Figure 5.10. The joints in the pipes must be carefully made and checked for watertightness as leakage would threaten the security of the dam. The staunching rings both hold the pipe in place and also prevent seepage along the pipe. They are cast directly against undisturbed earth, that is without using formwork. If the wall is to have a clay core, the pipe can pass through the core trench, with a concrete curtain wall as shown in Figure 5.10.

Clay core. The purpose of the clay core is to act as an impermeable barrier, and as long as it does this, the width does not matter and can

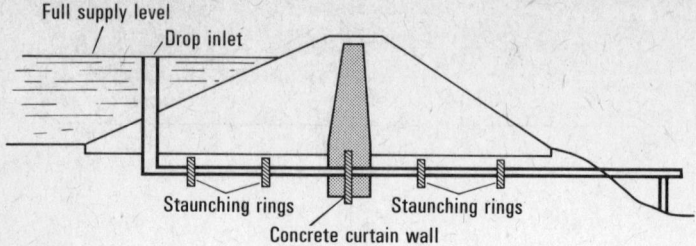

Figure 5.10.   Installation of an outlet pipe under a dam wall. (From Department of Conservation and Extension, Government of Rhodesia, *Handbook of basic instructions for dam construction.*)

be made to suit the method of construction. If the trench is excavated by hand, and the core material is placed by hand, the core can be as little as 1 m wide, but if done by machine it will have to be at least as wide as the machines. The excavated material from the trench can be used in the downstream side of the wall. The important points in constructing the core are as follows.

1. It must be taken down to, and bedded into, an impervious layer so there is no risk of seepage below it.
2. It must be continued up to the full supply level.
3. The material used for the core must have enough clay so that it is plastic and workable when wet. For dams which will remain full for most of the time, where the core will not dry out, a pure clay can be used, but for smaller dams which may dry up a pure clay might shrink and crack, so a sandy clay will be better.
4. The material must be properly placed and compacted. The best way is to place the clay in thin layers, add enough water to make it plastic, and to 'puddle' it, that is, to work it and compact it by labourers tramping it with their feet and ramming it with poles. Another good method is to drive cattle or oxen or buffaloes up and down over each layer of clay. When using earth-moving machinery the soil should be placed by a scraper in thin layers, and compacted when wet by wheeled tractors or by sheep-foot rollers. Placing the soil by bulldozer is not satisfactory, nor is compaction by tracked tractors.

Building the embankment. The rules for placing the soil of the embankment are the same as for the core, that is, it should be placed in thin layers, and well compacted. Adding water may assist the compaction, but this is a refinement not usually necessary on small dams. Placing soil by bulldozer is always undesirable, because it will not be properly compacted.

The cost of building the wall is a large part of the total cost of the dam, so it is important to move the earth as economically as possible. Earth-moving equipment is nearly always charged by the hour of working time so the important points are:

(1)   the equipment should be in good condition and capable of working at full efficiency;

(2)   the power of the tractor must be matched to the size of the earth carrier, so that the capacity of one is not limited by the capacity of the other;

(3)   the operation of the machinery must be carefully planned so that there is a minimum turn-round time for each cycle of load, carry, spread, and return.

Loading the scraper is easier when done downhill, and in hard ground it will help if the surface is loosened with a ripper or chisel plough. On bigger jobs when several units are employed, a great saving of loading time is achieved by push-loading. This is having one bulldozer in the borrow pit to push each scraper from behind while it is loading. Then the full scraper goes off and the pusher gets back in position for the next one.

An important development is the introduction in recent years of the 'self-loading' scraper. When a conventional scraper is half full the entering soil has to push back the soil already in the bowl of the scraper. This is the main reason why big scrapers need large crawler tractors to load them. The self-loading scraper has an endless chain elevator which lifts the newly entering soil over that which is already in the bowl. This allows the scraper to be operated by a smaller tractor, or to load more quickly, either of which means lower costs per cubic metre of earth moved.

The turn-round time of an earth-moving unit depends partly on the loading time, but mainly on the length of haul. For crawler tractors the time should be kept down to about 4 minutes under average conditions, or up to 6 minutes in difficult conditions. The borrow pit should be as close as possible to the wall, but not nearer than 10 m from the up-stream toe. The use of temporary ramps up on to the wall can reduce the turn-round time a great deal. On a long wall it may well be worth building several ramps which can be bulldozed into place, and levelled afterwards. Figure 5.11 shows how ramps and a figure-8 pattern can nearly double the rate of placing soil in the wall.

Where there is a variation in the soil available for the embankment,

(a)

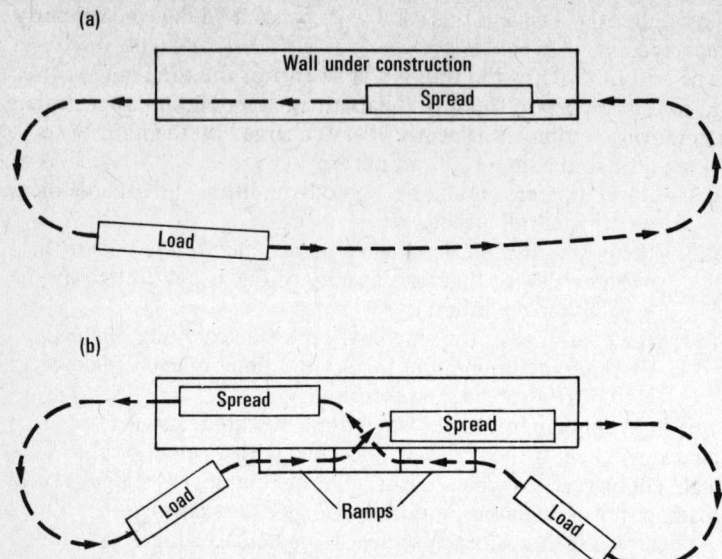

(b)

Figure 5.11.   The use of loading ramps can improve the efficiency when building a dam wall. (a) Simple but uneconomical pattern of earth-moving unit; (b) improved turn-round by using ramps.

any more porous material should be placed on the downstream side. Unsuitable soils were defined in Section 5.1.1. If the soil is sandy clay or clay it may not be necessary to have a clay core but in this case a 'hearting' should be made in the middle of the dam using selected material placed and compacted more carefully than the rest.

Control of side slopes. The tractor driver cannot tell how much to reduce the width as the wall rises and it is difficult to correct errors afterwards. To help the driver, stakes should be set out frequently showing the required width for the height reached (Figure 5.12). The stakes must be set out each time from the centre line, or the situation can arise as in Figure 5.12 (c) where the width of the wall is correct but the slopes are still not right.

Finishing works. The embankment should be given a thin cover of the topsoil removed from the site of the embankment and from the borrow pits, and then planted or sown with a suitable grass. The crest and down-

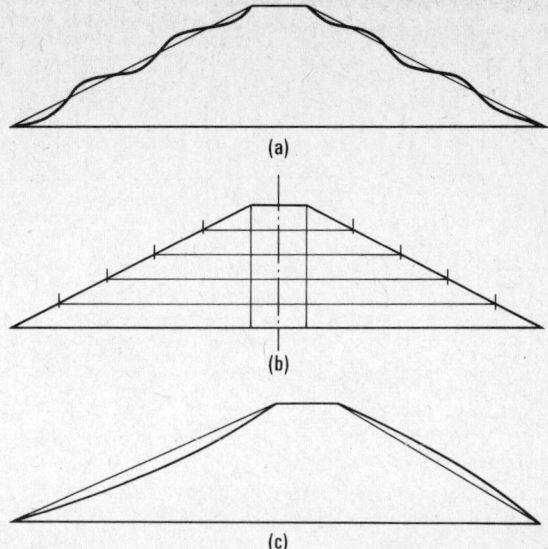

Figure 5.12.   Control of side slopes during construction: (a) without control the slopes will be uneven; (b) control stakes must be set out from the centre line; (c) correct width but not properly centred.

stream slope should have a good tough dryland creeping grass, and the upstream side one which will tolerate occasional drowning. The spillway and training wall should also be carefully grassed, and given other protection as required, such as stone pitching.

It is always sound practice to fence off the embankment and spillway, and if cattle will water at the dam it is preferable to fence off the whole dam and to pipe water to a drinking trough below the dam wall.

5.2. Small weirs

Weirs are structures of concrete, brick, or masonry, where the water flows over the top when the storage is full. They may be more suitable than earth dams in some situations, for example,

(1)   when there is a permanent flow which is too big to pass through a spillway;

(2)   when the object is to raise the water level so that part of the flow can be diverted into an irrigation canal, but storage is not required;

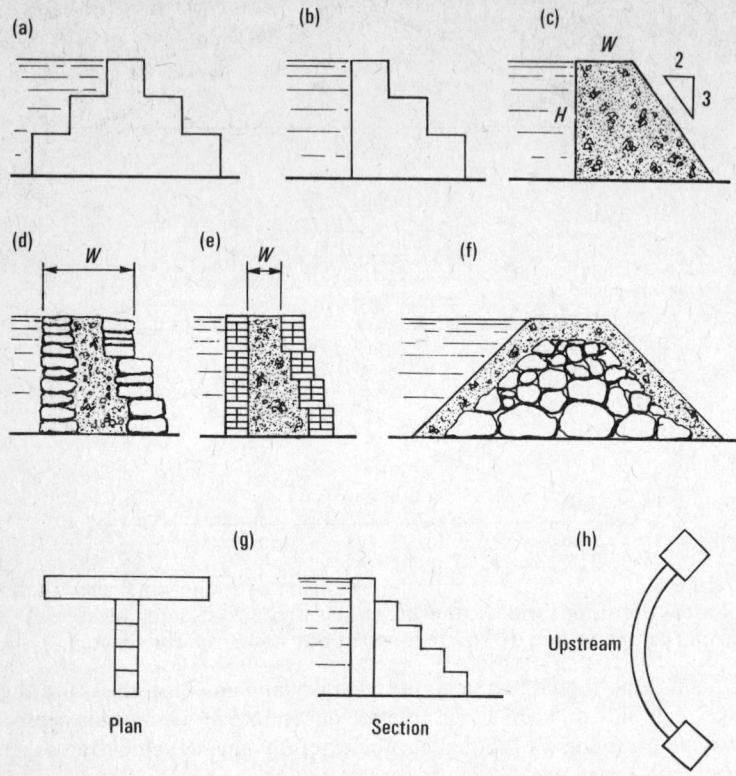

Figure 5.13.   Examples of small weirs. (a), (b), and (c) Sections through gravity weirs; (d) masonry used as shuttering for a concrete weir; (e) brickwork used as shuttering for a concrete weir; (f) a rock-fill dam with a concrete skin; (g) a slab and buttress weir; (h) an arch weir. (From Department of Conservation and Extension, Government of Rhodesia, *Handbook of basic instructions for dam construction.*)

(3)   when the object is gully control, or to trap sediment.

The main types of weir are shown in Figure 5.13. *Gravity weirs* depend solely on weight for their stability. A symetrical section (Figure 5.13 (a)) can be used, but the section in Figure 5.13 (b) gives the same height with less material and is more usual. For weirs not built in steps (Figure 5.13 (c)), the back slope is at 2 horizontal to 3 vertical, so the width W at any height h is given by

$$W = w + \tfrac{2}{3}(H-h),$$

where w is the crest width and
    H is the height of the weir.

The crest width depends on the wall height as follows:

    wall height up to 1 m, crest width 0·3m
    wall height 1 - 1·5 m, crest width 0·4 m
    wall height 1·5 - 2 m, crest width 0·5 m.

The weir can be entirely concrete, as Figure 5.13 (c), formed by pouring concrete into wooden shuttering. Alternatively masonry can be used to form the outside layer and also to act as the shuttering as in Figure 5.13 (d). Brickwork can also be used as shuttering as in Figure 5.13 (e), but in this case the brickwork is not counted as part of the structure and the dimensions for stability should be measured to the inside of the brickwork.

Another kind of gravity weir is the rock-fill dam (Figure 5.13 (f)), in which a pile of rocks and boulders gives the stability and a concrete skin makes it watertight.

The large amount of material necessary for gravity weirs can be reduced by *slab-and-buttress weirs* (Figure 5.13 (g)), or *arch weirs* (Figure 5.13 (h)). In the slab-and-buttress, the vertical slab is held up against the pressure of water by the buttress, and in the arch weir the pressure of water is resisted by the buttresses at each end. Both these kinds of weir can be built of concrete, brick, or masonry and can be built up to a height of about 1 m (3 ft) without design calculations. An engineer should be called in for larger weirs, or in cases where failure would cause serious damage.

Construction. The essential requirement for all weirs is a solid rock foundation extending the full width of the stream channel. Building weirs on poor foundations or on a rock bar which only goes half-way across may be possible, but should be left to engineers.

The first step is to thoroughly clean off all dirt and soil, chip off any loose or cracked rock, and if the surface is worn smooth by the water it should be roughened by chipping. To get a good bond between the foundation and the concrete a grout (cement and water mixed to the consistency of thick cream) is brushed onto the rock surface. Brick or masonry shuttering is built using a mortar mix of 6 sand to 1 cement (by volume) and all joints should be completely filled with mortar.

The concrete of a gravity weir does not require great strength, but this should lead to the use of an economical mix of good materials, rather than the use of poor materials. The cement should be fresh, and the sand should be clean and free from soil, dirt, and organic matter. River sand is quite suitable if clean. If not it should be washed. The coarse stone aggregate must be clean and free from soil and dirt. Ideally it should be a graded mixture of stones from 2 mm to 50 mm (0·08–2 in). River washed pebbles and gravel are suitable if clean.

A suitable mix is 1 part cement by volume, 3 parts sand, 6 parts stone. The quantities required per cubic metre of concrete are 0·15 m^3 cement, 0·5 m^3 sand, 1 m^3 stone, and 70 l of water. The order of mixing is not the same for mechanical mixers and mixing by hand. For mechanical mixers the stone is put in first, then the sand, then the cement, and then the water. For hand-mixing the cement and sand are thoroughly mixed dry, then the stone, and finally the water.

New concrete should not be allowed to dry out too quickly as this causes shrinkage, cracking, and loss of strength. It should be 'cured' for 7 days, that is, kept moist by covering with wet sacks. The most common causes of poor quality concrete are:

(1)   using sand or stone contaminated with soil;
(2)   using too much water, and
(3)   not curing properly.

5.3. Off-stream storage

In some situations it may be impractical to build storage works directly on the stream, and in this case the answer may be off-stream storage, that is, a storage reservoir which is filled by diverting water into it or by pumping. Off-stream storage is usually more expensive because it looses the advantage of the storage capacity in the stream channel, so it is most likely to be used for purposes which justify the higher cost, such as stock watering or irrigation. One example is the kind of irrigation storage known as a *balancing reservoir* or *night-storage reservoir*, which is used when there is a small but constant supply which has to be stored up for a while until there is enough for an irrigation application. For cattle watering, the advantage of off-stream storage is that it can be sited exactly where it is required—in a particular paddock, or near a bore-hole.

The cost of the water is mainly influenced by the ratio of the volume of storage capacity to the volume of excavated earth, known as the *storage/excavation ratio* or S/E ratio. Just digging a hole and storing water in it would give an S/E ratio of 1·0, but this can usually be increased by using the excavated soil to form a surrounding bank to contain water above original ground level. A naturally occuring depression may also lead to a good S/E ratio. Several shapes have been worked out which give the best S/E ratio for different conditions. When choosing the best size and shape it is necessary to consider the slope of the ground, the kind of water supply, and the use for the water.

Gently sloping ground from 1 per cent to 4 per cent will give better S/E ratios than flat ground, and filling with flood run-off by gravity will be cheaper than pumping from a stream or bore-hole. In low-rainfall areas, combined systems are common where catch drains are used to increase the catchment area, with a pumping system to top up when rainfall is low (Figure 5.14). In hot, arid climates, an important design consideration may be to minimize evaporation by having more depth and less surface area.

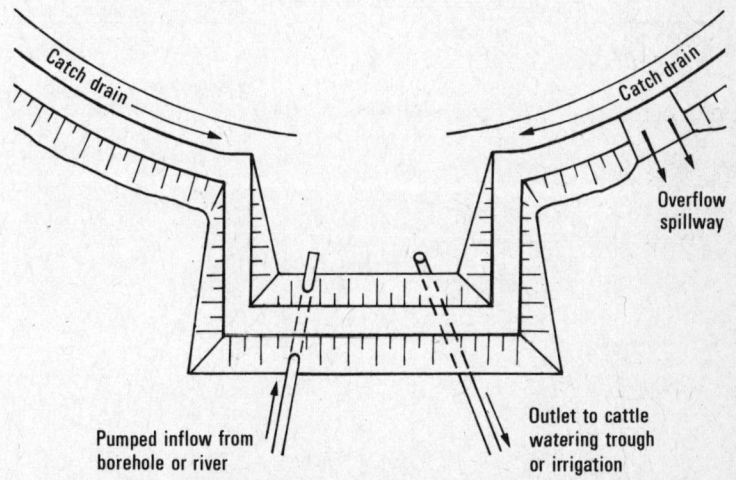

Figure 5.14.   Off-stream storage in a tank dam.

Seepage losses are likely to be more important than in the case of on-stream storage, and the higher value of the stored water makes it more important to consider how seepage may be reduced. The possible ways of reducing seepage, considered in order of increasing costs are as follows.

1. *Puddling the soil.* If there is sufficient clay in the soil, just working it when saturated may seal it.
2. *Additives.* An old dodge used successfully in South Africa for years is to add cattle manure to a leaky dam. There does not

appear to be any scientific explanation of how this works, but
there is no doubt that it often does.

A similar method is adding bentonite, which is a very fine clay. This
disperses, and seals up the small pore spaces. It is best added dry and
worked into the soil surface before filling the dam, but can also be
added to the water after filling.

3. *Floor-lining.* The most expensive treatment is to line the floor.
   This can be done with concrete, but this is expensive and liable to
   crack, or with bituminous or asphaltic compounds, or with
   membranes. Polythene or other plastic membranes are cheap, but

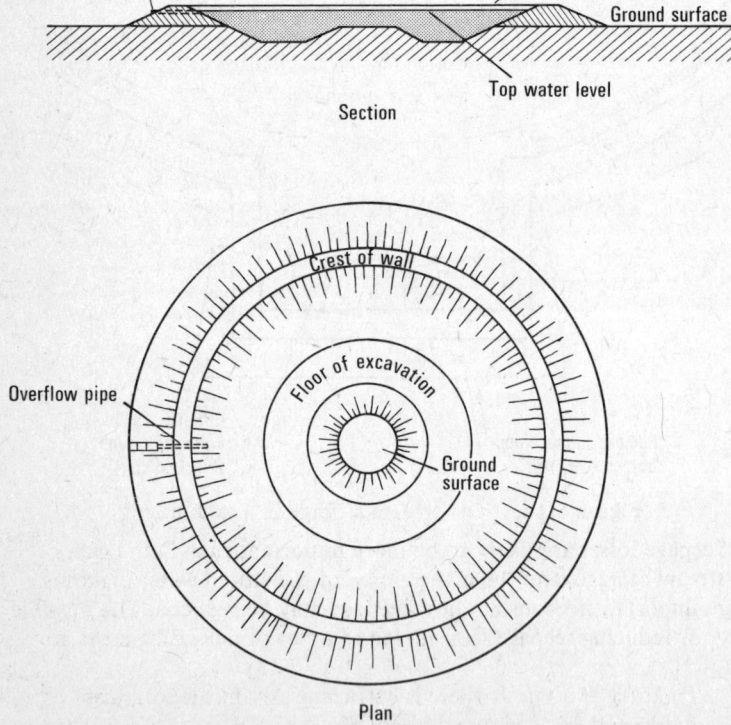

Figure 5.15.   Off-stream storage in a ring dam. (After Burton, *Water
storage on the farm*, W.R.F. Australia.)

need careful installation and may not last long. Butyl rubber liners are expensive, but reliable and long lasting.

The earth-moving equipment which can be used for construction of off-stream storage dams includes the towed scrapers used for earth dams, and also bulldozers and drag-line excavators.

Rectangular tanks. This is the best shape for gently sloping land and can give S/E ratios of 1·5 to 2·5 (Figure 5.14). *Catch drains* pick up surface run-off, and when the reservoir is full any surplus is spilled safely through overflows in the catch drains. These are also called *tank-dams*. Small sizes for stock watering can be built economically using either a drag-line excavator, or a small farm tractor and scoop.

Ring tanks. On flat, or nearly flat ground, better S/E ratios are obtained from round tanks rather than square tanks (Figure 5.15). Because of the geometry of this shape, the S/E ratio increases with size being about 1·5 for a 50 m diameter tank holding about 35 000 m^3 (7·7 million gall), to 4·5 for 200 m diameter holding 200 000 m^3 (45 million gall). gall).

Turkey's nest tank. The point of this type, pioneered in Australia, is that the water is stored above ground level (Figure 5.16), which may be

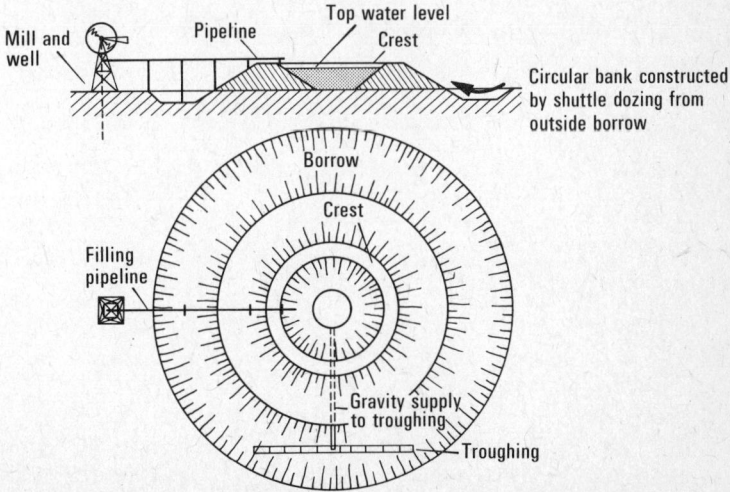

Figure 5.16.   Off-stream storage in a turkey's nest dam. (After Burton, *Water storage on the farm*, W.R.F. Australia.)

an advantage in either of two situations:

(1)   if the surface soil is relatively impermeable, but overlies a more porous material, or

(2)   where it is convenient to be able to distribute the water by gravity.

The S/E ratio is poor, from 0·2 at 200 m^3 (44000 gall) to 0·4 at about 5000 m^3 (1·1 million gall), and the main use is as an alternative to corrugated-iron storage tanks for small quantities of pumped water used for stock-watering.

6. Irrigation

6.1. Principles of irrigation

6.1.1. Opportunities and problems

There is not much new land waiting to be opened up for agriculture. The main increase in food supplies has to come from making better use of land already being farmed. More efficient crop production will help, for example better varieties, more use of fertilizer, mechanization, etc., but rainfall is still a limiting factor for a very large proportion of the earth's surface, and so irrigation is important in increasing production. Not that there is anything new about this. The development of civilized man was largely the result of irrigation which made possible bigger and more reliable yields of food crops. The history of irrigation goes back a very long time and there were major irrigation schemes four or five thousand years ago on the river Indus, the Nile, the Tigris, and the Euphrates, and on the rivers of China.

The big difference is that all these early schemes depended mainly on gravity, with sometimes a little man-power or animal-power, whereas today we can use large amounts of power to lift and pump water, and this means that a whole new opportunity for irrigation is opened up.

In addition to the obvious advantage of increased food supplies, irrigation has other potential benefits. Crop yields may be made more reliable, or two crops a year may be possible instead of one. Agriculture is still the main occupation in most developing countries, and if irrigation can make it 10 per cent more efficient this is usually a significant improvement to the economy. Irrigation can greatly increase the rate of generating capital from agricultural production. Multipurpose schemes can combine irrigation development with the generation of power for industrial development, with flood control, with population redistribution, with national policy (such as less dependence on imports), and with many other factors.

Types of irrigation. Three different types of irrigation can be identified.

1. In arid or semi-arid areas, where the soil is suitable but the rainfall is too low for crop production.

2. Where the rainfall is sufficient in quantity but badly distributed, so that the land alternates between uncontrolled flooding and drought.

Redistribution of the water may be achieved either by taking it from one part of the country to another, or by storing the surplus floods at one period for use during the dry season.

3. The rainfall may be usually sufficient in quantity, and reasonably distributed, but not reliable. Crop production in marginal areas can fluctuate greatly from year to year. Adding the little bit of extra water which is required at times when the rainfall is below average is called supplemental irrigation, because it is supplementing the rainfall. In terms of increased yield per cost of water applied, supplemental irrigation is usually very profitable.

Purposes of irrigation. Irrigation may benefit crop growth in several different ways, and the reason for irrigating may be any one of these, or a mixture of several.

1. To supply moisture which allows plant growth where there was none before, or to get better growth, or to extend the growing season.

2. As an insurance against drought.

3. To allow the movement of plant nutrients. The chemicals necessary for plant growth are absorbed by the plant in solution. For example, a surface application of a nitrogenous fertilizer will have no effect on a completely dry soil. There must be water to dissolve it and take it down to the roots, and if rain does not do this, perhaps irrigation can.

4. To leach undesirable salts, that is to wash them out in solution. In many parts of the world the build up of salinity is a most serious threat to agricultural production.

5. A less common purpose is to control the environment of growing plants. Irrigation can be used in different circumstances to prevent frost, to keep plants cool, or to keep them moist.

Problems. The problems likely to result from irrigation result from poor or incorrect practice. They are as follows.

1. *Salinity.* The build-up of saline salts in the soil can be from two causes. If the water table is raised by irrigation this can bring salts up into the root zone. The other way is from the application of irrigation water containing dissolved salts. In both cases the water is transpired by the plants and evaporated from the soil surface but the salts remain behind so there is a progressive build up over the years. The treatment of saline soils was discussed in Section 2.3.2.

2. *Waterlogging.* This can result from consistently applying too much water, or from inadequate drainage, or from seepage from a canal.

3. *Erosion and sedimentation.* Water flowing too fast in furrows or canals will cause soil erosion. The deposition of silt in canals is also undesirable. It may be difficult to avoid these entirely, but good design will solve most problems of this nature.

4. *Damage to soil structure.* Some kinds of sprinkler irrigation can damage the soil through excessive application rates, as discussed in Section 6.2.2.

6.1.2. Water requirements of crops

How much water does the crop need? Successful irrigation must start by trying to decide how much water is needed, and then supplying that need. To provide more water than necessary will be wasteful, but there must be enough for crop growth.

There are three ways to estimate the water requirement of crops.

1. We can grow the crop in an experimental tank called a lysimeter which allows us to measure the water applied, the amount drained out, and hence we know the amount evaporated. This is a useful research method but not suitable for the irrigation farmer.

2. The amount of water used by a crop depends on how much is evaporated from the leaves. This is affected by things like how hot and dry the air is and how much the wind blows. Another method is therefore to measure how much water is evaporated from an instrument designed for that purpose, and from this to estimate how much the crop would use during the same time. The instrument used is an evaporation pan, and this should be part of the equipment of any meteorological station in an agricultural area. It consists of a shallow metal pan raised slightly above the ground level and containing about 200 millimetres of water. A simple but workable pan can be cut from an old oil drum. At fixed times, such as daily or weekly, a reading is made of how much the level has dropped because of evaporation. A simply way to do this is to measure how much water must be added to bring the level back up to a fixed mark. Allowance must be made for rainfall into the pan.

The amount of water used by a crop is usually less than that evaporated from the pan, by an amount which varies during the plant's growth. Table 6.1 gives values for this ratio which is called the pan correlation factor.

3. Since the evaporation depends on climatic factors like temperature and humidity, the third method is to calculate an estimate of the crop use from measured meteorological data. Many formulas for doing this have been worked out, some simple, some complicated. Two simple

Table 6.1
Evaporation pan factors

The evaporation measured by the evaporation pan should be multiplied by the following factors to get an estimate of plant use:

Emergence to early growth	0·3 - 0·5
During vegetative growth	0·5 - 1·0
During flowering	1·0 - 0·8
During wet-fruit stage	0·8 - 0·6
During dry-fruit stage	0·6 - 0·0

From U.S.D.A. Soil Conservation Service Tech. Pub. 96.

Table 6.2
Values of k factor for estimating moisture requirements by the Blaney-Criddle method

		Low humidity	High humidity
Field crops			
	cotton	0·50	0·60
	maize	0·85	0·75
	sorghums	0·80	0·70
Orchard crops			
	citrus	0·50	-
	deciduous	0·70	0·60
Forage crops			
	pasture	0·90	0·70
	lucerne	0·85	0·75
Vegetable crops			
	peas	0·80	0·70
	beans	0·70	0·60
	potatoes	0·75	0·65

From U.S.D.A. Soil Conservation Service Tech. Pub. 96.

well-used methods will be described, both of which give useful estimates of the total crop use during each month. The *Blaney-Criddle method* was developed in California and is suitable for irrigation in arid climates:

$$U = k \frac{p \times t}{100} \; ,$$

where U is the monthly consumptive use in inches,

   p is the hours of daylight in the month expressed as a percentage of the annual hours of daylight (this will depend on latitude),

   t is the mean monthly temperature in $°F$, and

   k is the empirical constant which varies for the month and the crop. Some values of k are given in Table 6.2.

The *Thornthwate method* also depends on temperature and latitude, and is computed as follows.

1. Tabulate the mean monthly temperature $T°C$.
2. For each monthly value of T find the corresponding value of the heat index i from Table 6.3.

Table 6.3

Monthly heat index for estimating moisture requirements by the Thornthwaite method

$T (°C)$	i	$T (°C)$	i	$T (°C)$	i	$T (°C)$	i
1	0·1	11	3·3	21	8·8	31	15·8
2	0·3	12	3·8	22	9·4	32	16·6
3	0·5	13	4·3	23	10·1	33	17·4
4	0·7	14	4·8	24	10·8	34	18·2
5	1·0	15	5·3	25	11·4	35	19·0
6	1·3	16	5·8	26	12·1	36	19·9
7	1·7	17	6·4	27	12·9	37	20·7
8	2·0	18	7·0	28	13·6	38	21·6
9	2·4	19	7·6	29	14·3	39	22·4
10	2·9	20	8·2	30	15·1	40	23·3

From Zimmerman, *Irrigation*, Wiley, New York.

3. Add the twelve values of i to get the annual heat index I.
4. On Figure 6.1 draw a line joining the point of convergence to the appropriate value of I on the I scale. (The line drawn on Figure 6.1 takes $I = 125$ as an example.)
5. From the line drawn on Figure 6.1 find and tabulate, for each monthly value of T, the corresponding value of PET (*potential evapo-transpiration*). Note that at temperatures above $26·5°C$ the table at the bottom of Figure 6.1 is used instead of reading from the graph.
6. From Table 6.4 find and tabulate the monthly correction factors for the latitude. All latitudes more than $50°$ N or S use the factors

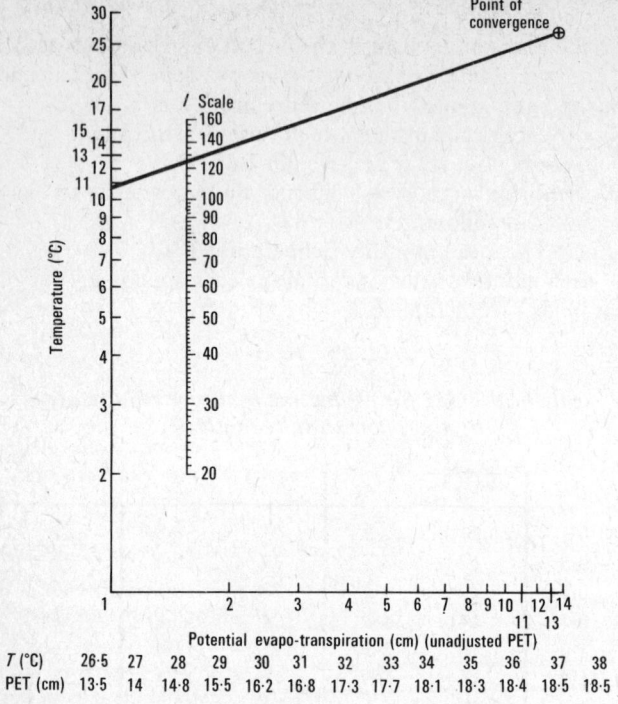

T (°C)	26·5	27	28	29	30	31	32	33	34	35	36	37	38
PET (cm)	13·5	14	14·8	15·5	16·2	16·8	17·3	17·7	18·1	18·3	18·4	18·5	18·5

Figure 6.1.   Nomograph for finding PET (potential evapo-transpiration) in the Thornthwaite method of estimating moisture requirement of crops. The table is used for temperatures above 26·5 °C. (After Zimmerman, *Irrigation*, Wiley, New York.)

for 50°. The monthly values of PET are then multiplied by the latitude correction factor to give the estimated monthly consumptive use of moisture. An example of the tabulation is shown in Table 6.5.

How much of the soil moisture is useful to plants?

In a completely saturated soil the large pore spaces are filled with what is called gravitational water because it can drain out under gravity. Because it drains away quickly this water is not much help to plant growth. The time taken for it to drain away varies for different soils, for sandy soils about one day, for clay soils up to 4 days. The moisture content remaining when free drainage stops is called *field capacity*.

Table 6.4
Latitude correction factors for the Thornthwaite method

N latitudes	Jan	Feb	Mar	Apr	May	Jun	Jul	Aug	Sep	Oct	Nov	Dec
0	1·04	0·94	1·04	1·01	1·04	1·01	1·04	1·04	1·01	1·04	1·01	1·04
5	1·02	0·93	1·03	1·02	1·06	1·03	1·06	1·05	1·01	1·03	0·99	1·02
10	1·00	0·91	1·03	1·03	1·08	1·06	1·08	1·07	1·02	1·02	0·98	0·99
15	0·97	0·91	1·03	1·04	1·11	1·08	1·12	1·08	1·02	1·01	0·95	0·97
20	0·95	0·90	1·03	1·05	1·13	1·11	1·14	1·11	1·02	1·00	0·93	0·94
25	0·93	0·89	1·03	1·06	1·15	1·14	1·17	1·12	1·02	0·99	0·91	0·91
26	0·92	0·88	1·03	1·06	1·15	1·15	1·17	1·12	1·02	0·99	0·91	0·91
27	0·92	0·88	1·03	1·07	1·16	1·15	1·18	1·13	1·02	0·99	0·90	0·90
28	0·91	0·88	1·03	1·07	1·16	1·16	1·18	1·13	1·02	0·98	0·90	0·90
29	0·91	0·87	1·03	1·07	1·17	1·16	1·19	1·13	1·03	0·98	0·90	0·89
30	0·90	0·87	1·03	1·08	1·18	1·17	1·20	1·14	1·03	0·98	0·89	0·88
31	0·90	0·87	1·03	1·08	1·18	1·18	1·20	1·14	1·03	0·98	0·89	0·88
32	0·89	0·86	1·03	1·08	1·19	1·19	1·21	1·15	1·03	0·98	0·88	0·87
33	0·88	0·86	1·03	1·09	1·19	1·20	1·22	1·15	1·03	0·97	0·88	0·86
34	0·88	0·85	1·03	1·09	1·20	1·20	1·22	1·16	1·03	0·97	0·87	0·86
35	0·87	0·85	1·03	1·09	1·21	1·21	1·23	1·16	1·03	0·97	0·86	0·85
36	0·87	0·85	1·03	1·10	1·21	1·22	1·24	1·16	1·03	0·97	0·86	0·84
37	0·86	0·84	1·03	1·10	1·22	1·23	1·25	1·17	1·03	0·97	0·85	0·83
38	0·85	0·84	1·03	1·10	1·23	1·24	1·25	1·17	1·04	0·96	0·84	0·83
39	0·85	0·84	1·03	1·11	1·23	1·24	1·26	1·18	1·04	0·96	0·84	0·82
40	0·84	0·83	1·03	1·11	1·24	1·25	1·27	1·18	1·04	0·96	0·83	0·81
41	0·83	0·83	1·03	1·11	1·25	1·26	1·27	1·19	1·04	0·96	0·82	0·80
42	0·82	0·83	1·03	1·12	1·26	1·27	1·28	1·19	1·04	0·95	0·82	0·79
43	0·81	0·82	1·02	1·12	1·26	1·28	1·29	1·20	1·04	0·95	0·81	0·77
44	0·81	0·82	1·02	1·13	1·27	1·29	1·30	1·20	1·04	0·95	0·80	0·76
45	0·80	0·81	1·02	1·13	1·28	1·29	1·31	1·21	1·04	0·94	0·79	0·75
46	0·79	0·81	1·02	1·13	1·29	1·31	1·32	1·22	1·04	0·94	0·79	0·74
47	0·77	0·80	1·02	1·14	1·30	1·32	1·33	1·22	1·04	0·93	0·78	0·73
48	0·76	0·80	1·02	1·14	1·31	1·33	1·34	1·23	1·05	0·93	0·77	0·72
49	0·75	0·79	1·02	1·14	1·32	1·34	1·35	1·24	1·05	0·93	0·76	0·71
50	0·74	0·78	1·02	1·15	1·33	1·36	1·37	1·25	1·06	0·92	0·76	0·70

S latitudes	Jan	Feb	Mar	Apr	May	Jun	Jul	Aug	Sep	Oct	Nov	Dec
5	1·06	0·95	1·04	1·00	1·02	0·99	1·02	1·03	1·00	1·05	1·03	1·06
10	1·08	0·97	1·05	0·99	1·01	0·96	1·00	1·01	1·00	1·06	1·05	1·10
15	1·12	0·98	1·05	0·98	0·98	0·94	0·97	1·00	1·00	1·07	1·07	1·12
20	1·14	1·00	1·05	0·97	0·96	0·91	0·95	0·99	1·00	1·08	1·09	1·15
25	1·17	1·01	1·05	0·96	0·94	0·88	0·93	0·98	1·00	1·10	1·11	1·18
30	1·20	1·03	1·06	0·95	0·92	0·85	0·90	0·96	1·00	1·12	1·14	1·21
35	1·23	1·04	1·06	0·94	0·89	0·82	0·87	0·94	1·00	1·13	1·17	1·25
40	1·27	1·06	1·07	0·93	0·86	0·78	0·84	0·92	1·00	1·15	1·20	1·29
42	1·28	1·07	1·07	0·92	0·85	0·76	0·82	0·92	1·00	1·16	1·22	1·31
44	1·30	1·08	1·07	0·92	0·83	0·74	0·81	0·91	0·99	1·17	1·23	1·33
46	1·32	1·10	1·07	0·91	0·82	0·72	0·79	0·90	0·99	1·17	1·25	1·35
48	1·34	1·11	1·08	0·90	0·80	0·70	0·76	0·89	0·99	1·18	1·27	1·37
50	1·37	1·12	1·08	0·89	0·77	0·67	0·74	0·88	0·99	1·19	1·29	1·41

From Zimmerman, *Irrigation*, Wiley, New York.

Table 6.5
*Example of calculation of consumptive use
by the Thornthwaite method*

Month	1 Mean monthly temperature (°C)	2 Heat Index i	3 PET (cm)	4 Latitude factor	5 Monthly consumptive use (cm)
Jan	16	5·8	2·4	1·00	2·4
Feb	28	13·6	14·8	0·91	13·5
Mar	32	16·6	17·3	1·03	17·8
Apr	32	16·6	17·3	1·03	17·8
May	30	15·1	16·2	1·08	17·5
Jun	28	13·6	14·8	1·06	15·7
Jul	24	10·8	9·5	1·08	10·3
Aug	24	10·8	9·5	1·07	10·2
Sep	22	9·4	7·0	1·02	7·1
Oct	22	9·4	7·0	1·02	7·1
Nov	22	9·4	7·0	0·98	6·9
Dec	18	7·0	3·6	0·99	3·6

Total 138·1 = I

Column 1 from meteorological records; column 2 from Table 6.3, column 3 from Figure 6.1; column 4 from Table 6.4 for latitude 10°N; column 5 from column 3 x column 4.

There still remains in the soil the moisture held loosely between the soil particles by surface tension. This is called capillary water, and is the main source of moisture for plant growth. If this is progressively taken up by plant roots the soil dries out until there is no more left and the plant has to stop growing—this is called *wilting point*. There will still be some moisture left in the soil but it is so tightly held onto the surface of soil particles that the plant roots cannot pull it off. This is called hygroscopic water or unavailable moisture.

The moisture available to plants is therefore that between field capacity and wilting point. The moisture content which corresponds to field capacity and to wilting point varies for different soil types, as shown in Figure 6.2. The available soil moisture is expressed as the

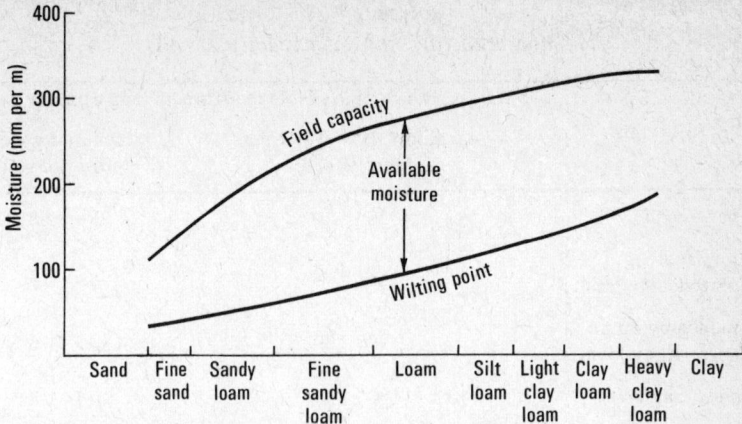

Figure 6.2. The available soil moisture in soils of different textures. (From U.S.D.A. Yearbook.)

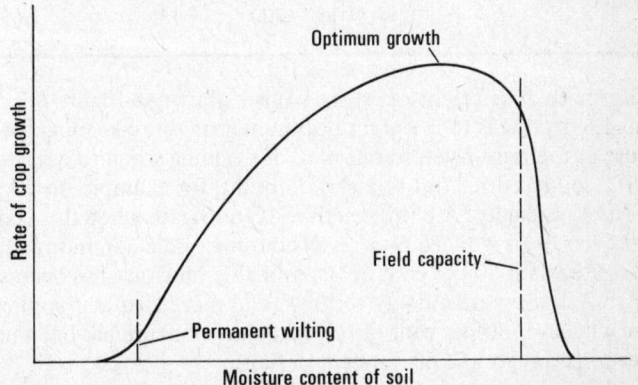

Figure 6.3. The effect of soil moisture on the rate of plant growth. (After Israelson, *Irrigation principles and practice*, Wiley, New York.)

water available per unit depth of soil, that is millimetres of water per metre of soil or inches of water per foot of soil, and the probable range of values is shown in Table 6.6.

When to irrigate? All of the available moisture is not equally available to the plants. When the soil moisture is close to field capacity the roots can take it up more easily than when it is near to wilting point. The

Table 6.6
Available moisture-holding capacity of soils

Soil type	Range of available moisture-holding capacity	
	(millimetres of water per metre of soil)	(inches of water per foot of soil)
Coarse sand	30 - 70	0·4 - 0·8
Fine sand Coarse loamy sand	60 - 100	0·7 - 1·2
Fine loamy sand Coarse sandy loam	90 - 130	1·1 - 1·6
Sandy clay loam	120 - 150	1·4 - 1·8
Silty clay loam	150 - 220	1·8 - 2·6
Clay loam	120 - 220	1·4 - 2·6
Sandy clay Clay	130 - 200	1·6 - 2·4

rate of growth at different moisture levels is shown in Figure 6.3. The object of irrigating is to maintain optimum growth by keeping the soil moisture at the right level, so one way of deciding when to irrigate is when the soil has dried out to a certain point, for example, to 70 per cent of field capacity. A better method is to irrigate when the available moisture has been reduced by a given amount, and a common practice is to irrigate when 50 per cent of the available moisture has been used. Sandy soils drain more quickly so they need more frequent applications, such as whenever 40 per cent of the available moisture has been used. Heavier soils can go to a 60 per cent deficit.

Deciding when to irrigate depends on the stage of crop growth as well as the soil moisture. During germination and very early stages of growth more frequent waterings are desirable, and it is particularly important to avoid stress due to a shortage of moisture at flowering and when the fruit or seed is swelling. On the other hand, it can sometimes be advantageous to deliberately keep the plant short of moisture, for example, to bring on the onset of flowering or to increase the sucrose content of sugar cane before harvesting.

It is sometimes useful to be able to predict when irrigation is likely to be required. This can be done by keeping a running budget of the

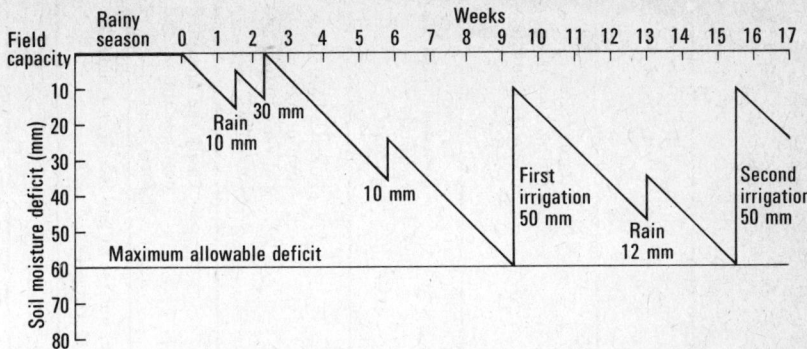

Figure 6.4.   A soil moisture budget to predict the need for irrigation.

soil moisture, as in Figure 6.4. Any moisture above field capacity is ignored, because it will drain away quickly. When there is no rain the soil moisture reduces as the soil dries out (the diagonal sloping lines). The moisture is brought up by any rain or irrigation. The moisture deficit at which irrigation is to be applied has been previously decided (for example 50 per cent of available moisture), and so it is easy to predict when this level will be reached. Further refinements can be added if a more accurate accounting is required, for example, if the consumptive use of moisture by the crop changes then a different draw-down diagonal slope can be used. The accounting can be done on a daily or weekly basis.

How to estimate the soil moisture? In order to know when to irrigate it is necessary to measure the soil moisture, and several methods are available. Using the appearance of the plants is not a good idea because the signs of drought, such as wilting and turning a bluish colour, only appear after the plant is short of water, whereas the object is to avoid this happening. The two most common instruments for measuring soil moisture are resistance blocks which measure the electrical resistance of the soil, and tensiometers which measure the soil-moisture tension. Direct measurement of soil moisture can be made by the gravimetric method, which consists of weighing soil samples before and after drying them in an oven. All these methods are useful for research, and are sometimes used on large schemes, but are not suitable for small-scale irrigation. The best method for farmers is to estimate the moisture by the feel of the soil. Using the data of Table 6.7, it is quite easy to estimate the available moisture with sufficient accuracy to be able to tell when to irrigate.

Table 6.7
Estimating soil moisture from the feel

Percent of the available moisture	Light soils Loamy sands and sandy loams	Medium soils Very fine sandy loam and silt loam	Heavy soils Silty clay loams and clay loams
0 - 25 (0 per cent is wilting point)	Dry, loose, flows through fingers	Powdery, sometimes slightly crusted, but easily broken down into a powdery condition	Hard, cracked; difficult to break down into powdery condition
25 - 50	Appears to be dry, will not form a ball with pressure	Somewhat crumbly, but holds together with pressure	Somewhat pliable, balls under pressure
50 - 75	Tends to ball under pressure, but seldom holds together when bounced in hand	Forms a ball, somewhat plastic, sticks slightly with pressure	Forms a ball, ribbons out between thumb and forefinger
75 - 100	Forms a weak ball, breaks easily when bounced in hand	Forms a very pliable ball	Easily ribbons out between thumb and forefinger
100 (Field capacity)	Upon squeezing, no free water appears on soil, but wet outline of ball is left on hand; soil sticks to thumb when rolled between thumb and forefinger		
Saturated	Free water appears on soil when squeezed		

How much to apply? If it has been decided to irrigate when the available moisture is down to 50 per cent then we should add enough water to bring it back to field capacity for the depth of soil containing the roots. There is no point in making it wetter than field capacity because the surplus would drain away, and there is no point in watering the soil at a depth where there are no roots. The depth that includes most of the roots for different crops is shown in Table 6.8.

The moisture is not drawn uniformly from the whole of the root zone and the approximate withdrawal pattern is shown in Figure 6.5.

Table 6.8
Effective rooting depth of some common crops

	mm
Fruit trees	750
Lucerne	1200
Cotton	900
Maize, small grains, wheat	600
Most vegetables	300

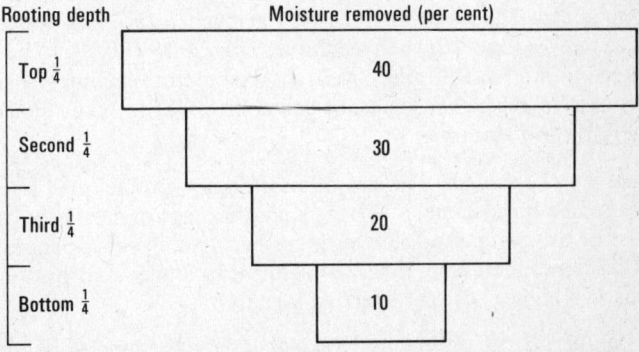

Figure 6.5. The extraction of soil moisture by plant roots.

6.1.3. Choice of irrigation method

When deciding on the best method of irrigation many factors must be considered. The things which are the deciding factors in one case may be unimportant in another. The following factors therefore cannot be listed in order of importance and in any one scheme not all of them would be relevant, but they comprise a list of all the main points which could influence the decision on how to irrigate.

National requirements. Particularly in the case of large schemes which use national or international capital, the national or political requirements may dominate. For example, large-scale plantation schemes may have the advantage that they are likely to bring in foreign investment but, on the other hand, the Government may wish to go for intensive small-holdings in order to increase the number of owner-farmers. The choice of crop may also be affected by national policy, for example, to diversify away from an over-dominant crop, or to encourage production of a crop in order to reduce imports.

Local traditions and skills. In many countries there are traditional irrigation methods with deeply inbred skills in their operation. It might be better to continue the traditional method thàn to introduce some new method which is theoretically better but not liked or understood by the people. Changing from flood irrigation to sprinkler is a case in point. On the other hand, new methods can be acceptable if they become symbols of status and progress. The introduction of machines can be as important because of the boost to the image of farming, as because of the improved performance.

Mechanization. The traditional reason for mechanization in North America and Europe was to save labour. This is more likely to be a disadvantage in most developing countries. The use of machinery may still be desirable to get higher yields, or better crops, or to take some of the drudgery out of farming.

The scale of the project. The size of an irrigation project may influence the costs. In a big scheme, involving a large storage dam, it is better if the cost of this can be spread over more hectares, or more people. On the other hand, increasing the size might mean longer distribution canals and a higher cost per irrigated hectare.

Water factors. If the water supply is higher than the land to be irrigated, surface irrigation will probably be easiest. If the water has to be pumped up this will increase the chance of sprinklers being better. If the water supply is far away the choice of conveyance is linked to the method: pipes go with sprinkler systems; furrows go with surface methods. The amount of water and rate of flow is a big factor. Surface irrigation uses more water (up to 15 per cent more) and also needs large flows.

As to the water quality, contamination by sediment indicates surface application, because it interferes with pumps, pipes, and sprinklers. Salinity in the water also suggests flood irrigation so that the salts can be flushed out.

The higher the cost of water the greater is the required efficiency. High-price water is therefore more often associated with sprinkler irrigation. Where water is expensive and land is cheap, the measure of profitability of irrigation may be the yield per amount of water applied, not the yield per hectare.

Soil factors.

1. *Topography*

Nearly level slopes are suitable for any method.
Gentle slopes are best for furrow or border irrigation.
Steeper slopes suggest sprinkler.
Regular uniform slopes are needed for borders.

2. *Soil type*

Soils with a high storage capacity (that is, large available moisture content) suggest surface applications, with large amounts applied infrequently. Soils with low storage require more frequent applications of smaller amounts, and this suggests sprinklers.

Soils with medium infiltration rates are suitable for any method, but soils with low infiltration (less than 10 mm/hour or 0·5 in/hour) are best for flood irrigation, and soils with high infiltration (more than 80 mm/hour or 3 in/hour) are easiest to irrigate by sprinkler. Light sandy soils liable to erosion are unsuitable for furrow irrigation, so are saline soils.

The depth of soil may determine whether levelling for surface application is possible.

Crop factors.

1. Sprinkler irrigation is best for shallow-rooting crops or any other crops requiring light, frequent waterings.

2. For row crops, furrows are best, sprinklers are possible, and borders are unsuitable.

3. For forage crops, borders are best.

4. For orchard crops, furrows are best.

5. Tall crops like maize or sugar cane make sprinkler irrigation more complicated because they need tall risers to get the sprinklers up near the top of the crop.

6. Some diseases like mildews and viruses may be spread by the splash of sprinkler irrigation. Soil splash may also cause unacceptable contamination of table crops such as strawberries or lettuce.

Climatic factors. Where strong winds are probable, sprinkler irrigation is less suitable because the distribution will be uneven, and hot dry winds will increase the evaporation loss. For flood irrigation the evaporation loss is only about 2 per cent, but it can be up to 15 per cent with sprinklers.

Frequency of applications. The question is whether the crop can go for 10 days or 2 days between waterings. This depends on the soil moisture storage, and the crop requirement, and whether the irrigation is supplementing rainfall. In general, sprinkler irrigation has more flexibility for frequent, light waterings, and for supplementary irrigation. The frequency will have a big effect on costs. Increasing the frequency of application will increase the labour requirement in the case of flood irrigation, and the amount of equipment in the case of sprinklers.

Table 6.9
Items of capital and recurrent costs of irrigation

	Surface schemes	Sprinkler irrigation
Capital costs	Water supply Control works and supply canals Distribution canals Land-levelling Border checks Drains and outlets Bridges over canals and drains	Water supply Pumps and engines Piping and valves Sprinklers
Recurrent costs	Maintenance of canals, etc. Re-levelling Labour	Pumping costs Depreciation of pumps, piping, etc. Labour

Economic factors. The main items of capital expenditure and recurrent costs for both surface and sprinkler irrigation are given in Table 6.9. It is not possible to generalize about the relative costs because individual items can vary so much from one scheme to another. For example, land-levelling might be a very large expense in one surface scheme and negligible in another.

6.2. Sprinkler irrigation

6.2.1. When to use sprinklers

Sprinklers have the following advantages.

1. Levelling is not required, so sprinklers can be used on hilly or uneven land unsuitable for flood irrigation. Soils can be irrigated which are too shallow for levelling, and there is no risk of fertile topsoil being buried during the levelling.
2. Light, frequent waterings are possible, as required on shallow soils, and for germination and plant establishment.
3. Uniform distribution of water is possible.
4. The amount of water is easily controlled.
5. There are no permanent field obstacles such as ditches.
6. Small in-flows of water into the system can be used.
7. The equipment can be mobile, and this is a vital requirement for supplementary irrigation.

Sprinklers have the following disadvantages.

1. The efficiency will be reduced by strong winds or low humidity.
2. The capital cost of pumps, pipes, and equipment is high.
3. Sprinkler irrigation needs a regular water supply. It cannot use flood run-off, nor distribution schemes where water is supplied to the farmers in turn.

6.2.2. Overhead irrigation methods

Let us look first at the different kinds of sprinklers and sprays, and then at the different field systems for using them. The operating pressures and intensity of application are tabulated in Table 6.10.

Sprinklers and sprays.

1. *Reaction rotation sprinkler* (Figure 6.6). This is the kind most often used for lawns and in private gardens. The rotating head is driven round by the reaction of the jets at the end of two or four arms. The jets are adjustable to give a fine or coarse spray, and it can operate at almost any pressure. The distribution is fairly even but it can only cover a small circle. It is very cheap, but has little field application.

2. *Fixed-head sprays* (Figure 6.6). These are used for lawns and orchards. The vertical jet strikes the downward pointing cone and is spread out in a flat cone spray. The advantages are that it is robust and has no moving parts. It can put down fairly high intensities, but gives a poor distribution.

3. *Nozzle lines* (Figure 6.6). Nozzles are fixed at intervals along a pipe which is oscillated through an arc of about 90 - 120° by a water-

Table 6.10
Pressures and application rates of various types of sprinkler

Type	Purpose	Pressure kN/m²	kgf/cm²	lbf/in²	Application rate mm/hour	in/hour
Reaction spray	Lawns		any			
Fixed head	Lawns or orchards	35-550	0·35-5·5	5- 80	15-20	0·6 -0·8
Nozzle lines	Vegetables and market gardening	175-350	1·75-3·5	25- 50	5 maximum	0·2
Perforated pipe		70-350	0·7 -3·5	10- 50	10 minimum	0·4
Slow rotation Sprinklers						
Low pressure	When higher pressures not possible	70-200	0·7 -2·0	10- 30	3-25	0·15-1·0
Medium pressure	General agricultural use	200-400	2·0 -4·0	30- 60	5-45	0·25-1·8
High pressure	Grass or forage crops	400-700	4·0 -7·0	60-100	5-50	0·25-2·0

An explanation and conversion for units are given in the Appendix.

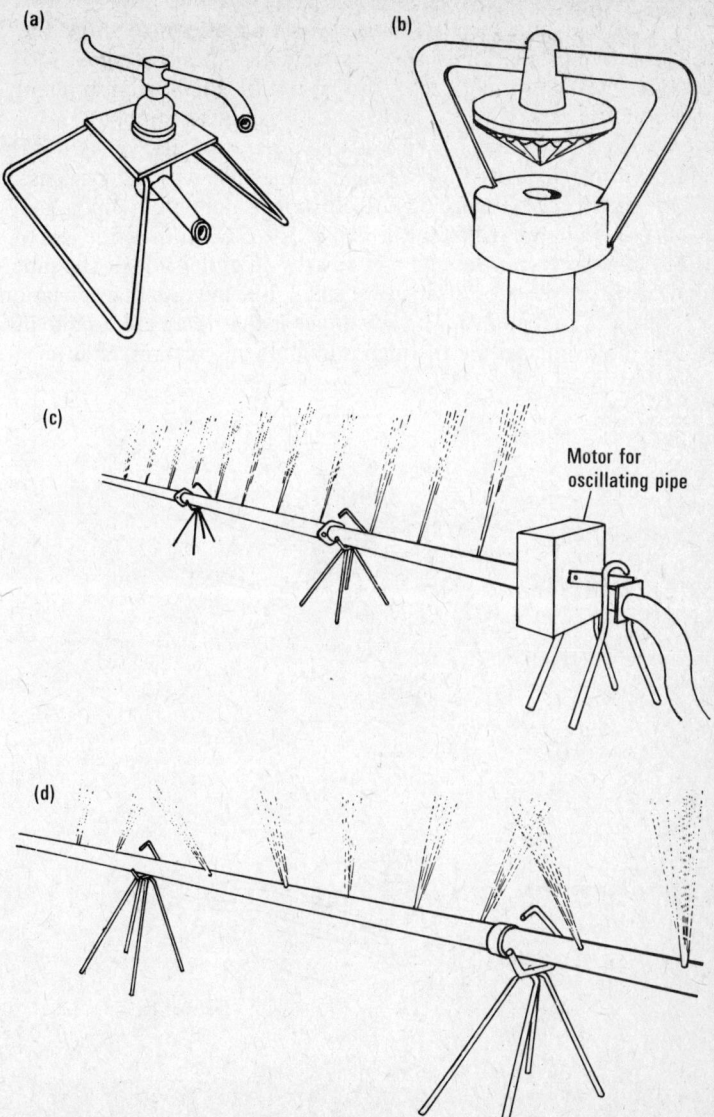

Figure 6.6.   Some types of irrigation sprinkler equipment: (a) reaction-rotation sprinkler; (b) fixed-head spray; (c) nozzle lines; (d) perforated pipe.

driven motor. The pipe lines are usually fixed a metre or so above the ground, and the installation is often permanent. The advantage is low application rates with small size of drops, and the application is mainly in nurseries. The capital cost is high, and a clean water supply is required to avoid blocking of the small nozzles. A smaller version of the same basic principle is used for watering domestic lawns and gardens.

4. *Perforated pipe* (Figure 6.6). Lightweight aluminium piping has many very small holes drilled in the top of the pipe so that fine jets come out at different angles and wet an area on either side of the pipe. Like nozzle lines, the drop size is very small, but the rate of application is much higher (Table 6.10). The advantage is that there are no moving parts. The disadvantages are the high minimum application, and the distortion by wind.

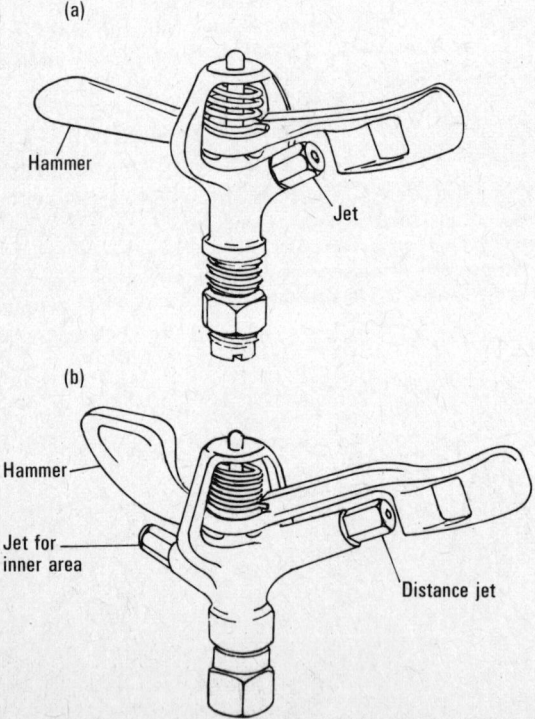

Figure 6.7.   Slow-rotation irrigation sprinklers: (a) single nozzle; (b) double nozzle.

Table 6.11
Example of manufacturers sprinkler data

(a) Metric units

Nozzle diameter (mm)	Pressure† (bars or kgf/cm^2)	Radius of throw (m)	Flow (m^3/hour)	Precipitation (mm/hour) for spacings (m)			
				18 × 18	18 × 24	24 × 24	24 × 27
4·0	2·5	14·0	0·94	2·90	-	-	-
	3·0	14·5	1·02	3·15	-	-	-
	3·5	15·0	1·09	3·36	2·52‡	-	-
5·0	2·5	15·5	1·52	4·69	3·52‡	-	-
	3·0	16·0	1·67	5·15	3·87‡	-	-
	3·5	16·5	1·80	5·55	4·17	3·13‡	-
6·0	2·5	17·0	2·20	6·79	5·09	3·82‡	-
	3·0	17·5	2·44	7·53	5·65	4·24‡	-
	3·5	18·0	2·60	8·02	6·02	4·51‡	4·01‡
7·0	2·5	18·0	2·84	8·76	6·57	4·93‡	4·38‡
	3·0	18·5	3·11	9·60	7·20	5·40	4·80‡
	3·5	19·0	3·45	10·65	7·97	5·99	5·32‡
	4·0	19·5	3·76	11·60	8·70	6·53	5·80‡
8·0	2·5	18·5	3·75	11·57	8·68	6·51	5·79‡
	3·0	19·5	4·16	12·84	9·63	7·22	6·42‡
	3·5	20·5	4·50	13·88	10·42	7·81	6·94
	4·0	21·5	4·98	15·37	11·53	8·65	7·68
10·0	3·0	21·0	6·81	-	15·76	11·82	10·51
	3·5	22·0	7·12	-	16·48	12·36	10·99
	4·0	23·0	7·64	-	17·68	13·26	11·79
	4·5	24·0	8·14	-	18·84	14·13	12·56

The data above is for Model SR15 of Irrifrance Ltd.

† The SI unit for pressure is the newton per square metre, but manufacturers of pumps, piping, and irrigation equipment are more likely to use the bar (atmospheric pressure), millibars, or the metric unit of kilograms force per square centimetre (kgf/cm²). An explanation and conversions are given in the Appendix.
For practical purposes, 1 bar = 1 kgf/cm² = 100 kN/m² = the head of 10 m of water.

‡ For a triangular positioning.

(b) English units

Operating pressure (lb/in^2)	Nozzle sizes (in)												
	5/32 × 1/8		11/64 × 1/8		3/16 × 1/8		13/64 × 1/8		7/32 × 1/8	
	Dia (ft)	Discharge (gall/min)	Dia (ft)	Discharge (gall/min)	Dia (ft)	Discharge (gall/min)	Dia (ft)	Discharge (gall/min)	Dia (ft)	Discharge (gall/min)
30	86	5·2	89	5·8	92	6·6	95	7·4	98	8·3
35	87	5·6	91	6·3	95	7·1	97	8·0	100	8·8
40	88	6·0	93	6·8	97	7·6	100	8·5	103	9·4
45	90	6·3	94	7·2	98	8·0	102	9·0	104	10·0
50	92	6·7	96	7·5	100	8·4	103	9·5	106	10·6
55	93	7·0	98	7·9	102	8·8	104	10·0	108	11·1
60	94	7·3	99	8·3	103	9·3	106	11·4	109	11·6

This data is for S.P.P. model STN 75 M5.

Slow-rotation sprinklers. This is by far the most important type, and is used for the great majority of field installations. Their design and manufacture is highly competitive, with manufacturers in many countries offering a wide range of sprinklers for every use and purpose. Most sprinklers have either a single nozzle (Figure 6.7(a)) or two opposing nozzles (Figure 6.7 (b)). With two nozzles, one sprays the inner area, and the other, sometimes called the distance jet, extends the diameter of application. The nozzles are interchangeable and available in different sizes to suit different pressures as shown in Table 6.11.

The usual method of rotating the sprinklers is in a series of small jerks from the impact of a swinging arm. The jet of water hits an angled plate and swings it to one side against a spring. As soon as it moves out of the jet the spring returns it and the impact against a stop moves the sprinkler round a few degrees, and the cycle repeats. On larger machines an impeller is rotated by the water jet, and this drive goes through reducing gears to give a slow rotary movement to the whole sprinkler. There are sprinklers with devices to limit the rotation to part of the circle, for use at the end of fields or in corners. Many special-purpose sprinklers are made such as those with a flat, low trajectory for operating under fruit trees, or with large nozzles for spraying livestock effluent, or for very high or very low pressures, and so on.

It is difficult to design a sprinkler which gives uniform distribution over the whole of the wetted circle. A typical distribution is shown in Figure 6.8, which also shows how the water application can be evened out by making the sprinkler patterns overlap.

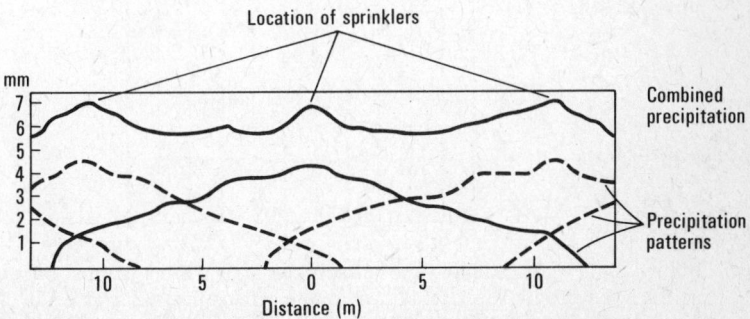

Figure 6.8.   The distribution pattern of precipitation from sprinklers. (From *Planned irrigation*, Wright Rain Ltd, Ringwood, Hants.)

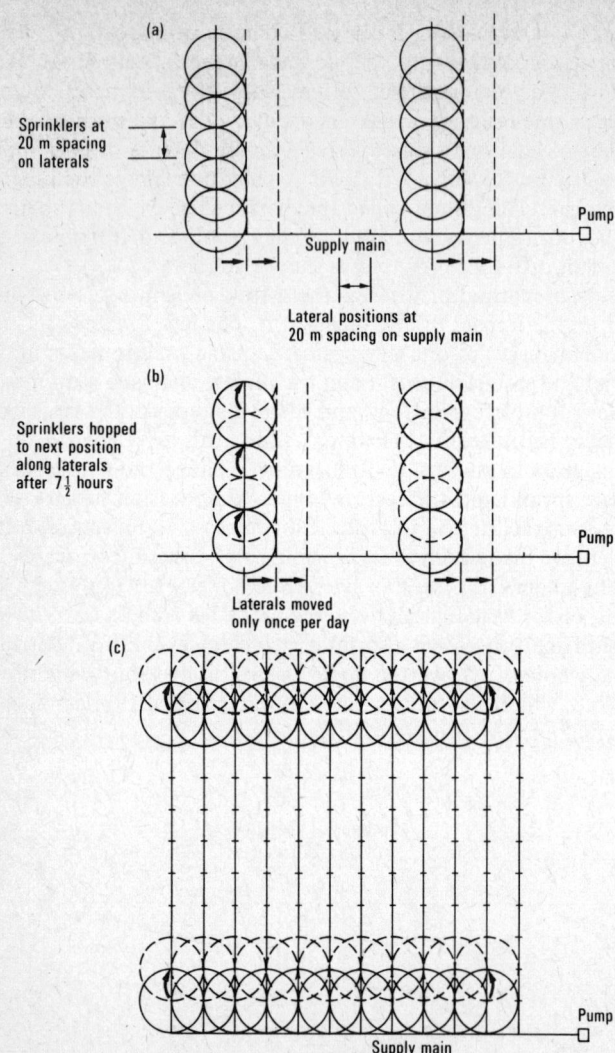

Figure 6.9.   Some hand-move sprinkler irrigation systems: (a) conventional system—the laterals, complete with a sprinkler at each position, are moved after each application of 5 or 6 hours; (b) Wright Rain 'Hopalong' system—each lateral has sprinklers at alternate positions; (c) Wright Rain 'Portagrid' system—two sprinklers on each lateral are moved progressively along the lateral.

Sprinkler systems. Sprinkler systems either use simple equipment and a lot of labour to move them, or more sophisticated equipment which reduces the labour required. The economics of which is better will depend mainly on the availability and cost of labour.

Hand-move systems are the simplest and have the least capital cost. There are two, three, or four laterals, which are the pipes on which the sprinklers are fixed (Figure 6.9). Each lateral has perhaps 10 or 15 sprinklers and remains in place for several hours before it is all moved to the next position.

The labour of moving the laterals can be reduced in systems such as the Wright Rain 'Hop-a-long' and the Farrow 'Risermatic'. These use more laterals, but only half the number of sprinklers on each lateral. The sprinklers are moved along the laterals periodically, and this operation is simplified by automatic self-closing valves on the laterals at each sprinkler position. This means that the more labour-consuming job of moving the laterals only has to be done once each day.

The next stage in labour-saving is to instal a full network of laterals. The cost may be reduced by using small-bore piping, although we shall see later that this may mean higher pumping costs. The Wright Rain 'Portagrid' uses 30 mm aluminium laterals along which a small number of sprinklers are progressively moved. The Farrow 'Cropset' is similar but uses small-bore plastic piping.

When sufficient piping and sprinklers are installed so that none of it has to be moved, this is called a *solid-set system*. Completely automatic operation can be achieved on solid-set systems by either automatic timers which switch valves on and off, or by 'sequencing valves' which automatically close after a given time and pass the water along to the next lateral.

The movement of laterals and their sprinklers is laborious, so *roll-move systems* aim to make this simpler. In the side-roll method (Figure 6.10 (a)) the pipe is emptied of water to reduce the weight, and then moved sideways on large wheels. The power is either a hand-cranking system or a small petrol engine, and the drive may be transmitted either directly along the pipe using extra-strong thick-walled piping, or by a separate drive shaft. The disadvantages are that the wheels can get bogged down in wet patches, and also the pipe has to be stopped with all the sprinklers vertical.

Another way of moving the lateral is the *trail-line*, which drags it longitudinally on wheels or skids (Figure 6.10 (b)). On some systems the wheels can be set at a slight angle to give the required progressive movement down the field.

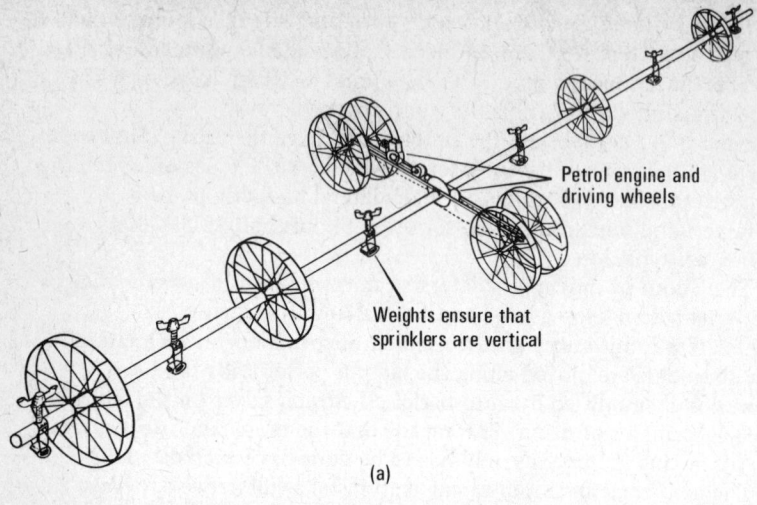

Petrol engine and
driving wheels

Weights ensure that
sprinklers are vertical

(a)

(b)

Figure 6.10.   Roll-move irrigation systems: (a) side-roll system; (b)
trail-move system.

Centre-pivot systems are large permanent systems which irrigate a
circular field by rotating a pipe line around a central point, which is
often next to the bore-hole supplying the water (Figure 6.11). Power is
provided by electric motors or water-powered turbine motors at several

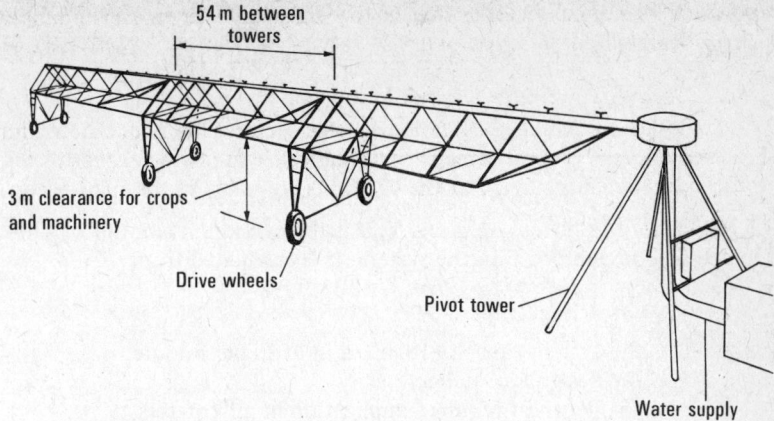

54 m between
towers

3 m clearance for crops
and machinery

Drive wheels

Pivot tower

Water supply

Figure 6.11.   The centre-pivot irrigation system.

places along the line, and the line moves slowly and constantly. An
example of this type is the S.P.P. Ranger series which can irrigate
circular fields up to 100 ha.

Another self-moving system has a long cable laid out from a sprinkler
to an anchor at the far end of the field and a long flexible pipe supplying the water. A water-driven turbine slowly winds up the cable on a
drum so drawing the sprinkler slowly across the field as in the Farrow
'Rainamatic' (Figure 6.12). Similar multi-sprinkler machines may be

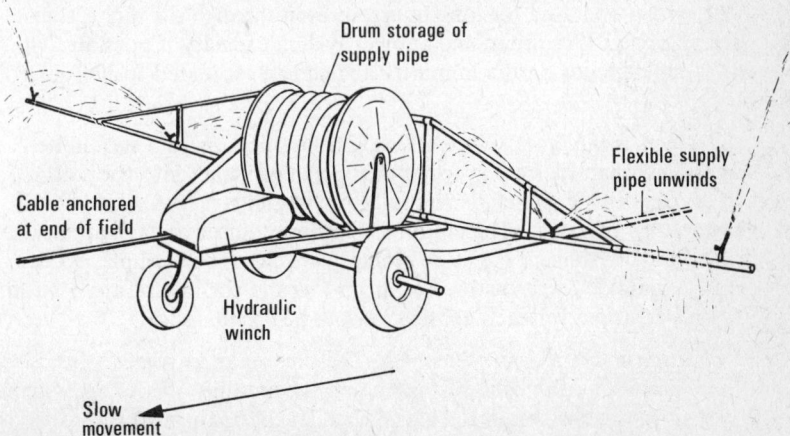

Drum storage of
supply pipe

Flexible supply
pipe unwinds

Cable anchored
at end of field

Hydraulic
winch

Slow
movement

Figure 6.12.   A self-moving irrigation system.

powered by a small diesel engine, as the 'Irri-anglia', or tractor-mounted as the 'Wright Rain Laureau'.

6.2.3. Design of sprinkler systems

To design an overhead irrigation scheme requires many decisions and choices, and the process is made much easier by following a step-to-step procedure which ensures that the decisions are taken in the right order.

1. *Calculate the system capacity.* This is the maximum amount of water which will be pumped into the system. It is calculated from

$$Q = 167 \times \frac{Hm}{fh} \times \frac{100}{e} \quad ,$$

where Q is the system capacity required in litres per minute,
   H is the design area in hectares,
   m is the depth of required application in millimetres,
   f is the frequency of irrigation in days,
   h is the number of hours of operation per day, and
   e is the efficiency of sprinkling (per cent).
In English units,

$$Q = 378 \times \frac{Ad}{fh} \times \frac{100}{e} \quad ,$$

where Q is in gallons per minute, A in acres, and d in inches. Frequency, hours of operation, and efficiency have the same units as in the metric equation.

The hours of operation must be decided at the outset. If the equipment can be operated for long hours, or even through the night, then less equipment is required and a lower system capacity is possible. The efficiency depends on the humidity and can be estimated from Table 6.12.

2. *Determine the rate of application.* The application rate should be slightly less than the rate at which water can infiltrate into the soil surface, and this depends on the soil texture and structure. Table 6.13 gives a range of values. The amount of water required at each application is known from the soil moisture storage capacity (for example, 50 per cent of available soil moisture), so from this and the rate of application the time required for each application can be calculated.

3. *Choose the sprinkler and spacing.* The precipitation rate is a function of the discharge from each sprinkler, and the amount of overlap, which depends on the spacing of the sprinklers. The basic formula is

$$Q = \frac{S_1 \times S_2 \times P}{3600} \quad ,$$

Table 6.12
Estimated sprinkler efficiencies

Amount of water applied during each irrigation (mm)	Cool temperatures Rate of application (mm/hour)			Moderate temperatures Rate of application (mm/hour)			Hot temperatures Rate of application (mm/hour)		
	less than 6	6 to 10	more than 10	less than 8	8 to 12	more than 12	less than 10	10 to 15	more than 15
More than 50	75	80	85	70	75	80	65	70	75
25 - 50	70	75	80	65	70	75	60	65	70
Less than 25	65	70	75	60	65	70	-	-	-

From *Planned irrigation*, Wright Rain Ltd., Ringwood, Hants.

These figures are for a dry climate, and can be increased by 5 per cent for humid climates.

Table 6.13
Maximum application rates for different soils

Soil texture	Maximum rate of application	
	mm/hour	in/hour
Coarse sand	20 - 40	0·75 - 1·5
Fine sand	12 - 25	0·5 - 1·0
Sandy loam	12	0·5
Silt loam	10	0·4
Clay loam Clay	5 - 8	0·2 - 0·3

The above rates are for level or gently sloping soils, with good structure or with cover. For soils with poor structure, or bare soils, reduce rates by 20 per cent. For soils steeper than 5 per cent, reduce rates by 20 per cent.

where Q is the discharge from each sprinkler in litres per second,

S_1 is the spacing of sprinklers along the laterals in metres,

S_2 is the distance between laterals in metres, and

P is the required precipitation in millimetres per hour.

In English units,

$$Q \text{ (gall/hour)} = \frac{S_1 \text{ (ft)} \times S_2 \text{ (ft)} \times P \text{ (in/hour)}}{115 \cdot 5}$$

The discharge from each sprinkler depends on the pressure and the size of the nozzles, and can be obtained from the manufacturers' catalogues. Examples are given in Table 6.11 and Figure 6.13.

The sprinklers are spaced so that the distribution is improved by overlapping as discussed in Section 6.2.2. The uniformity of distribution is improved by close spacing, but this increases the cost of the equipment. It is more difficult to get uniform distribution at low application rates, especially below 5 mm/hour.

The spacing may be square, in which case $S_1 = S_2$ and will be between 30 per cent and 65 per cent of the diameter wetted by each sprinkler. In rectangular spacings S_1 (along the lateral) is usually less than S_2 (between laterals), for example, 18 m × 24 m (or 40 ft × 60 ft). Triangular arrangements are possible, but less popular. In windy conditions the amount of overlap is increased, as shown in Table 6.14.

Selecting a suitable combination of sprinkler type, nozzle size, operating pressure, and spacing is made simpler than it sounds by the

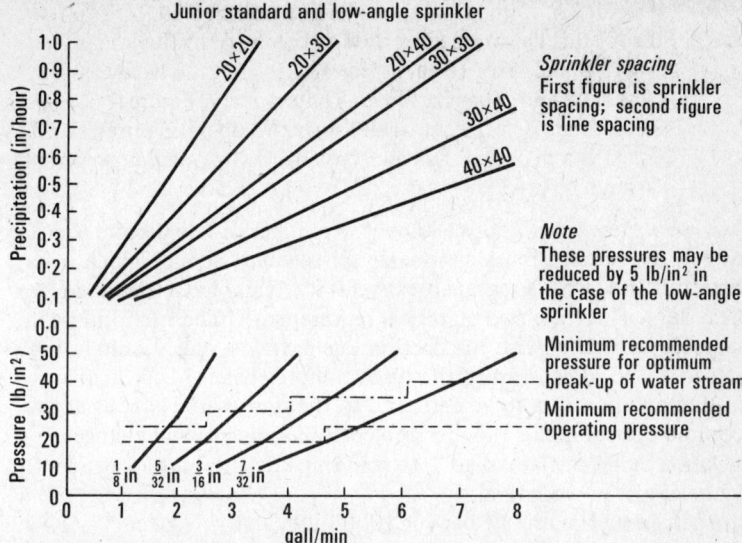

Figure 6.13. Manufacturer's data on sprinklers in graphical form. (From *Planned irrigation*, Wright Rain Ltd., Ringwood, Hants.)

Table 6.14
Spacing of sprinklers along laterals

Wind conditions	Sprinkler spacing on lateral (percentage of wetted diameter)
No wind	65
Up to 8 km/hour	60
8 - 15 km/hour	50
More than 15 km/hour	30

From *Planned irrigation*, Wright Rain Ltd., Ringwood, Hants.

detailed information supplied in the sprinkler manufacturers design sheets.

4. *Design the laterals.* If the pipes used for the laterals are too big they will cost too much and be heavy to handle; if too small the friction loss will be high and a bigger pump will be required. Good design is selecting the best size. The procedure is

(a)   Decide the number of sprinklers on each lateral.
(b)   From this determine the flow at the entry to the lateral.
(c)   By trial and error select a pipe size so that the head loss for this flow and pipe size (from Table 4.3 and Figure 4.6) is not more than 20 per cent of the sprinkler operating pressure. If there is a marginal choice between two sizes the *larger* should always be used.

5. *Design the main pipes.* We know the number of laterals and the flow in each, so we can choose a pipe size for the main supply which can carry this flow with a reasonable head loss. Again it is a trial-and-error procedure. There are two empirical requirements. The friction loss in the mains, together with the friction loss in the laterals should not exceed 25 or 30 per cent of the total pumping head. Also the mains must allow the water to be delivered to the start of the laterals at the required pressure, plus the pressure corresponding to any change of elevation along the lateral and the standpipe height, plus 75 per cent of the head loss in the lateral.

   Note. In metric units 1 bar (or 1000 millibars) = 1kgf/cm^2 = 10 m head of water.
   In English units to convert lb/in^2 to feet of head multiply by 2·31; to convert feet of head to lb/in^2 multiply by 0·43. See the Appendix for a more detailed discussion of units of pressure.

6. *Choose the pump and power unit.* Finally, as both the quantity and the total pumping head are now known, the pump and power unit can be chosen. Pumps and motors are discussed in Section 4.4.

6.3. Surface irrigation

6.3.1. When to use surface irrigation

The advantages of surface irrigation compared with sprinkler irrigation are:

(1)   the large expenditure on storage, headworks, and distribution canals is usually paid for by the Government, and the capital cost on the farm is low—this is an advantage in the case of peasant agriculture;

(2)   surface irrigation is more likely to be traditionally understood, and where there is no previous experience of irrigation it is easily taught;

(3) surface irrigation is more suitable for some crops, such as rice and forage crops;
(4) it is easier to use leaching to help avoid or control salinity problems;
(5) surface methods can use large flows available for short periods, and so allows a canalized water supply to be shared by several land-owners.

Some possible disadvantages are:

(1) it is difficult to get even distribution of water on light permeable soils which have high infiltration rates;
(2) surface irrigation is not suitable for crops which need frequent light waterings;
(3) efficient surface irrigation usually requires smooth land—if the land is not naturally smooth it may be expensive to level it, and topsoil and fertility can be disturbed;
(4) the field layout for surface irrigation may restrict mechanization unless special measures are adopted.

The question of whether to use surface or overhead irrigation thus depends on so many factors that it is not possible to lay down rules about when either will be the best. Every case will be different, and each must be considered separately taking account of all the conditions. If surface irrigation appears to be required, the question of choosing the most suitable of the different surface methods is relatively simple because it is mainly determined by the kind of crop to be grown. We can divide irrigation methods into two main groups—where the whole of the ground surface is to be uniformly flooded, and where the crop is grown in rows and the water is applied in furrows.

6.3.2. Flooding methods for field crops

Flooding is best for grass or forage crops, and for any other crop which is grown on the flat over the whole ground surface. It can be done very crudely and simply or with varying degrees of skill and care. *Wild flooding* (also called *contour flooding* or *water spreading*) is the simplest way, and consists of spilling water out of a supply ditch and allowing the water to flow freely down the slope. Unless the land is very smooth the water application is uneven, but it can be a very cheap simple method suitable for grass or forage crops where unevenness of growth does not matter. It is best on lands steeper than 2 per cent, and some smoothing of the bumps and hollows is usually worth while.

One application of this method is in making some use of storm run-off in semi-arid regions. A simple diversion weir or dam diverts the

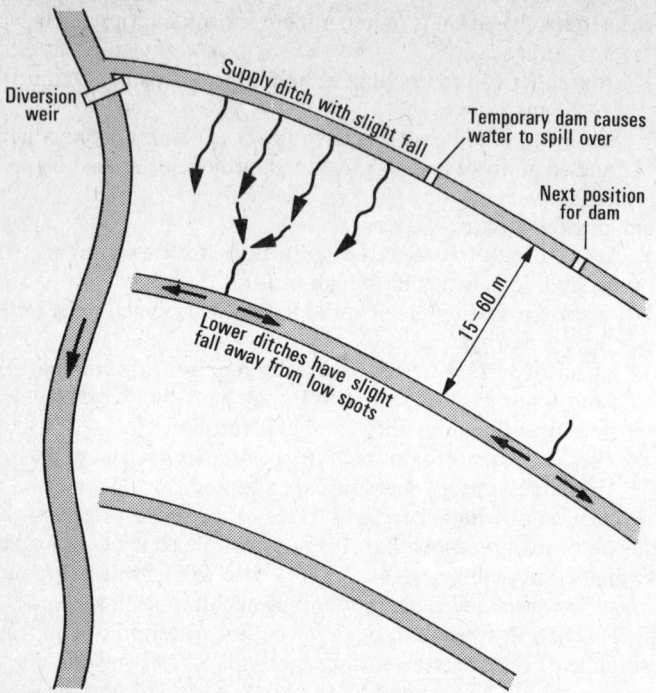

Figure 6.14.   Wild flooding is the simplest type of surface irrigation.

flood water and spills it onto any suitable land, often along the stream banks (Figure 6.14). Even if the efficiency of using the water is very low it is better than letting the storm run-off go completely to waste.

If the water supply is more regular it will be worth putting more effort into the scheme, for instance by constructing the supply ditch carefully on a grade of 0·5 per cent so that the water can be spilled out at regular intervals. The soil excavated from the ditch should be put on the uphill side, so that the water can spill smoothly over the downhill side. As the water flows downhill it tends to collect into streams, so other ditches can be built lower down the slope to redistribute it more evenly. Typical spacing of these interceptor ditches is from 15 m to 60 m (50 - 200 ft) down the slope. They are either on a level grade or built to slope away from the low spots where water collects.

If a continuous flow is being distributed the required flow will be about 1 l/s per hectare (about 5 or 6 gall/min per acre), but wild flooding is usually done with bigger intermittent flows.

The advantages of the method are that it is cheap and simple, and it requires little labour to operate and little or no land preparation.

Basin irrigation and *check flooding* are two similar kinds of flood irrigation, the difference being mainly in size. Check flooding is running large streams of water into large nearly level plots surrounded by earth banks which retain the water on the plot. It is suitable for grain or forage crops where large irrigation streams are available and the land is gently sloping.

The earth banks (called levees, or bunds, or dikes, or ridges) are built along a level contour at vertical intervals of 60 - 120 mm (2 - 5 in). At suitable intervals other banks at right-angles form the enclosed checks which may be up to several hectares on gentle slopes.

Basin irrigation also consists of running water into a level area surrounded by a bank, but is usually on a smaller scale, with basins from a few square metres upwards. The size and shape depend on the soil type, the size of irrigation stream, the crop, and the land slope. Suggested sizes are shown in Table 6.15. Soils with high infiltration

Table 6.15
Size of irrigation basins

Rate of flow	Area of basin (for each unit of rate of flow)			
	Sand	Sandy loam	Clay loam	Clay
100 l/s	0·07 ha	0·20 ha	0·40 ha	0·70 ha
100 m³/hour	0·02 ha	0·06 ha	0·12 ha	0·20 ha
1 ft³/s	0·05 acre	0·15 acre	0·30 acre	0·50 acre
1000 gall/min	0·13 acre	0·40 acre	0·80 acre	1·30 acre

From F.A.O. Handbook, *Surface irrigation.*

rates need smaller basins or the water will not be uniformly spread. The larger the irrigation stream, the bigger the basin which can be supplied. The slope affects the amount of earth-moving required to construct the basins, and the steeper the slope, the smaller is the optimum size of basin. One disadvantage of basins is that the banks and the supply ditches interfere with cultivation and harvesting. The layout which gives the widest spacing between supply ditches is to have long basins on either side of each ditch as shown in Figure 6.15.

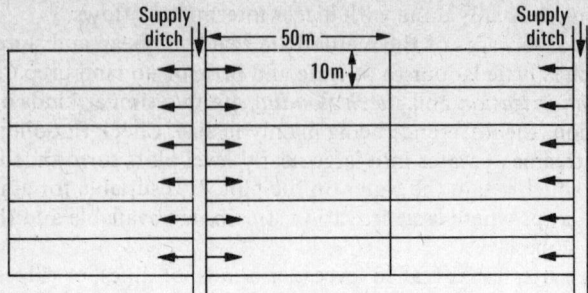

Figure 6.15.   A good layout of basin irrigation.

Basin flooding can be also used for orchards, either small circular or square basins for each tree, or low narrow basins with a line of trees. (See also furrow irrigation, pp. 153 - 162.)

Rice paddies and *rice terraces* are the forms of basin irrigation used for growing rice because it has different water requirements from most other crops. During most of its growing season rice is grown in standing water, the depth of which varies with the age of the plants. It would be difficult to achieve the necessary control of the depth of water by intermittent applications, so the usual method is to have the basins arranged like a very shallow flight of stairs. Water is led in to the top terrace, flows very slowly and gently through the crop, and overflows down to the next terrace, and in turn through all the terraces. Each basin or terrace is almost level and the water movement is controlled by the in-flow and out-flow. There has to be provision for draining the terraces before harvest.

Border strip or *strip check* irrigation is the most precise form of surface irrigation. It is particularly suitable on land with a gentle uniform slope and is best for close-growing crops like pasture, forage, and small grains. The land is very accurately levelled in long rectangular strips from 3 m to 20 m (10 - 60 ft) wide and up to 200 m (600 ft) long. A uniform grade runs the length of each plot, usually between 0·2 per cent and 1 per cent, but steeper grades can be used for perennial crops like pasture. The boundaries are raised earth banks like the banks around basins. Water is supplied from a canal to the top end of the strip and because of the even grading it flows in a regular uniform sheet down the strip. The water is cut off after a time which is calculated so that the water already on the strip continues down the strip and gives a uniform application. A simple rule is to shut off when the water has reached four-fifths of the length. The water is applied at the top end by syphons or turn outs from a supply canal above ground level. The labour requirement for border-strip irrigation is very low, and the efficiency high, but only if the land has been carefully and accurately levelled and smoothed so that the water spreads uniformly. The length and width of the strips depends on soil type, slope, and crop water requirement. Suggested combinations are given in Table 6.16.

The strips can be very long to suit mechanized handling of the crop if portable pipes are used to supply water at more than one point down the length of the strip. Light plastic pipes which fold flat and roll up after use are often used for getting water to the intermediate points.

The purpose of the borders is only to guide the flow of water down the strip, not to contain it like the banks round basins, so the borders are less substantial. The borders must be built up before the strip is levelled, otherwise there will be low spots where soil has been borrowed for the borders.

The size of the irrigation stream must be matched to the size and slope of the strip. If the stream is too small the result is an over-watering. This is the opposite of what one might expect at first sight, but the reason is that a small stream will advance slowly down the strip and result in too long an application. A bigger stream moves quickly down the strip, is cut off sooner, and gives a smaller total application. The first 10 - 15 m (30 - 50 ft) of the strip is often made level to help spread the water over the full width before it starts to flow down the strip.

6.3.3. Furrow methods for row crops

For any crop grown in rows it is more efficient to apply the water down furrows between the rows. An equal amount of water is released down each furrow from a slightly elevated canal at the head of the field. A recent development is the use of gated pipes, that is, pipes with small controllable outlets at spacings equal to the distance between furrows. Lightweight aluminium pipes started this trend, but are being replaced by collapsible thin-walled plastic pipes which roll up when not in use. Gated pipes allow the long runs which are essential to efficient mechanical crop-handling.

The advantages of furrows compared with flooding are:
(1) evaporation is less than when the whole surface is flooded;
(2) there is less danger of puddling on heavy soils;
(3) it is possible to cultivate sooner after irrigation;
(4) the method is suitable for a wide range of soils and slopes.

Table 6.16
Size of border strips

(a) Metric units

Soil type		Amount of each irrigation (mm)	Land slope (per cent)	Rate of flow per metre width (l/s)	Border strip	
Texture	Infiltration (mm/hour)				Width (m)	Length (m)
Sand	More than 25	100	0·2 - 0·4 0·4 - 0·6 0·6 - 1·0	10 - 15 8 - 10 5 - 8	12 - 30 9 - 12 6 - 9	60 - 90 60 - 90 75
Loamy sand	18 - 25	125	0·2 - 0·4 0·4 - 0·6 0·6 - 1·0	7 - 10 5 - 8 3 - 6	12 - 30 9 - 12 6 - 9	75 - 150 75 - 150 75
Sandy loam	12 - 18	150	0·2 - 0·4 0·4 - 0·6 0·6 - 1·0	5 - 7 4 - 6 2 - 4	12 - 30 6 - 12 6	90 - 250 90 - 180 90
Clay loam	6 - 12	175	0·2 - 0·4 0·4 - 0·6 0·6 - 1·0	3 - 4 2 - 3 1 - 2	12 - 30 6 - 12 6	180 - 300 90 - 180 90
Clay	0·25 - 6	200	0·2 - 0·3	2 - 4	12 - 30	350+

(b) English units

Soil type		Amount of each irrigation (in)	Land slope (per cent)	Rate of flow per foot width (ft³/s)	Border strip	
Texture	Infiltration (in/hour)				Width (ft)	Length (ft)
Sand	More than 1	4	0·2 - 0·4 0·4 - 0·6 0·6 - 1·0	0·11 - 0·16 0·09 - 0·11 0·06 - 0·09	40 - 100 30 - 40 20 - 30	200 - 300 200 - 300 250
Loamy sand	0·75 - 1	5	0·2 - 0·4 0·4 - 0·6 0·6 - 1·0	0·07 - 0·11 0·06 - 0·09 0·03 - 0·06	40 - 100 25 - 40 25	250 - 500 250 - 500 250
Sandy loam	0·5 - 0·75	6	0·2 - 0·4 0·4 - 0·6 0·6 - 1·0	0·06 - 0·08 0·04 - 0·07 0·02 - 0·04	40 - 100 20 - 40 20	300 - 800 300 - 600 300
Clay loam	0·25 - 0·5	7	0·2 - 0·4 0·4 - 0·6 0·6 - 1·0	0·03 - 0·04 0·02 - 0·03 0·01 - 0·02	40 - 100 20 - 40 20	600 - 1000 300 - 600 300
Clay	0·10 - 0·25	8	0·2 - 0·3	0·02 - 0·04	40 - 100	1200+

From F.A.O. Handbook, *Surface irrigation.*

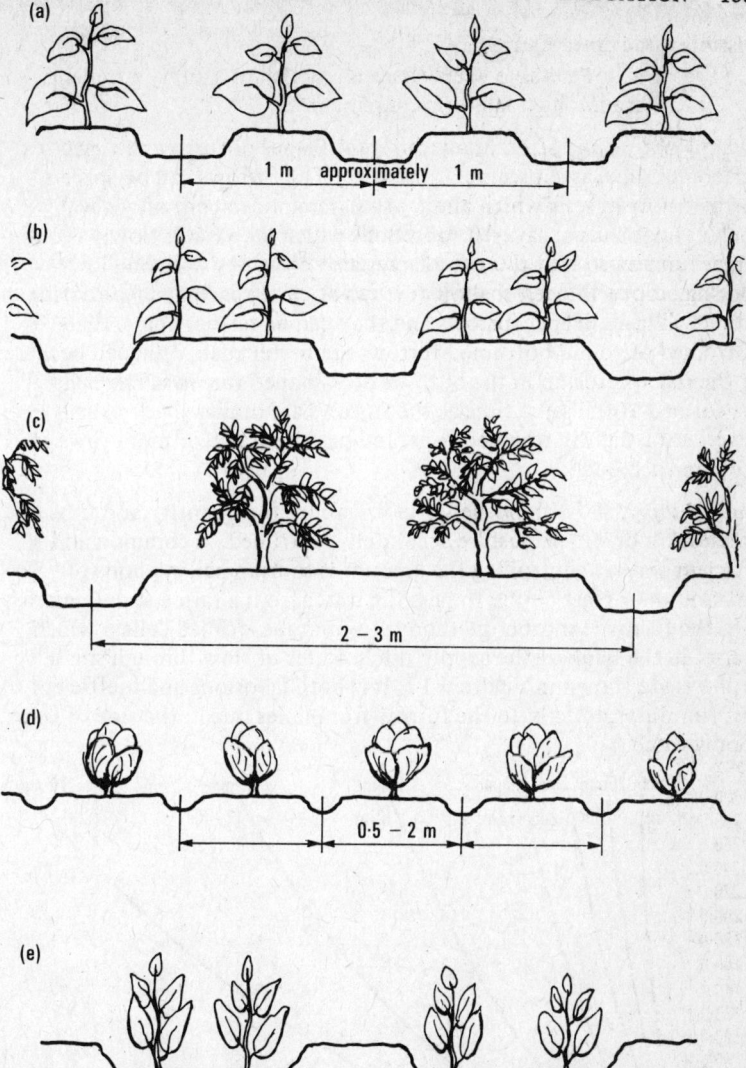

Figure 6.16.   Some different kinds of furrow irrigation: (a) single-row cropping; (b) double rows on each bed; (c) large beds for orchards; (d) small furrows (corrugations) for vegetable crops; (e) some crops are planted in the furrow instead of in the bed.

Possible disadvantages are:
  (1)   it is not suitable when there is any risk of salinity problems;
  (2)   there is a high labour requirement.

Furrow shapes and sizes. Many different shapes of furrow are used for
different crops, as shown in Figure 6.16. The furrows can be spaced
farther apart in soils which allow easy lateral movement of the water,
such as light soils or layered soils. Soils which take water slowly require
larger furrows so that there is a larger area of wetted soil. Shallow-
rooting crops will need shallow furrows at close spacing; deep-rooting
crops will need deeper furrows and they can be farther apart. Flat-
bottomed or round-bottomed furrows are better than V-shaped because
of the risk of erosion in the bottom of V-shaped furrows. The beds
are formed at the same time as the furrows are drawn. They usually
have flat or slightly rounded tops, and may have one or more rows of
crop on each bed.

Water supply. As with border-strip irrigation, the quantity and time of
application of water must be accurately controlled. A common and
efficient way of controlling the amount is to use small syphons of
aluminium or plastic tube to pass the water from a raised supply ditch
into the furrow. Another method is to use tubes (called spiles) which
are set in the bank of the supply ditch. Rates of flow through small
syphons are shown in Figure 6.17. It is both laborious and inefficient to
just run water directly to the furrow from holes dug in the side of the
supply ditch.

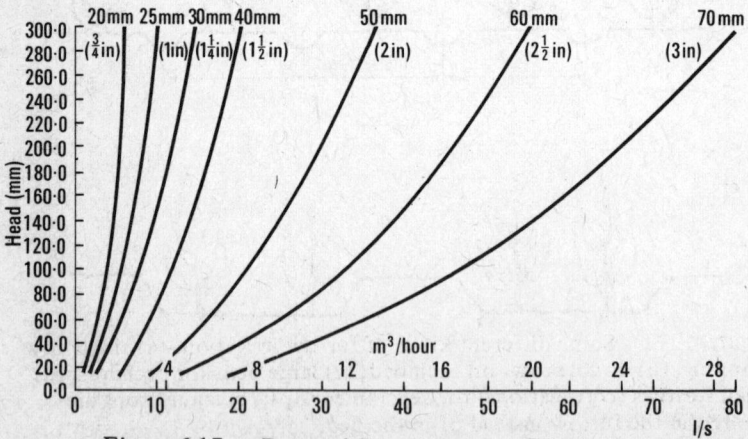

Figure 6.17.   Rates of flow through small siphons.

Uniform distribution of water down the whole length of the furrow can be achieved by adjusting the rate of flow. A larger flow is used to get the water quickly down the furrow, because if the water moves down too slowly the top end is saturated before water reaches the bottom end. As the soil becomes wet the rate of infiltration usually slows down, so that when the water has reached some fixed point, usually four-fifths of the length of the furrow, the flow is reduced to a quarter, and this lower flow is continued until sufficient water is applied.

Table 6.17
Application rates for furrow irrigation

Furrow slope (per cent)	Maximum rate of flow (l/s)	(gall/min)	Comments
0·1	6	100	Rate of flow limited by size of furrow.
0·3	2	33	Upper limit of slope where furrows can flow full without causing erosion.
0·5	1·2	20	Rate must be less than furrow capacity or erosion will occur.
2·0	0·3	5	Maximum non-erosive rate is much less than furrow capacity.
More than 2·0	-	-	Not recommended because of risk of erosion.

From F.A.O. Handbook, *Surface irrigation.*

Some general guides to maximum rates of application are shown in Table 6.17, but efficient furrow irrigation is essentially based on experience gained by trial and error. Holes should be dug at the top, middle, and bottom of the furrow to see how much soil has been wetted, and the time or quantity adjusted accordingly. A useful rule is that the water should reach the end of the furrow in one-quarter of the total application time.

With short furrows, or soils where the infiltration rate is much less after the initial wetting, it may be possible to use only one rate of flow, cutting it off when the wetted front has reached a point fixed by trial and error.

It is not possible to arrange the water application to give a uniform distribution down the furrow without there being some wastage from the bottom end of the furrow, and there must be provision for handling this water. The best thing is to collect it for re-use, either by pumping

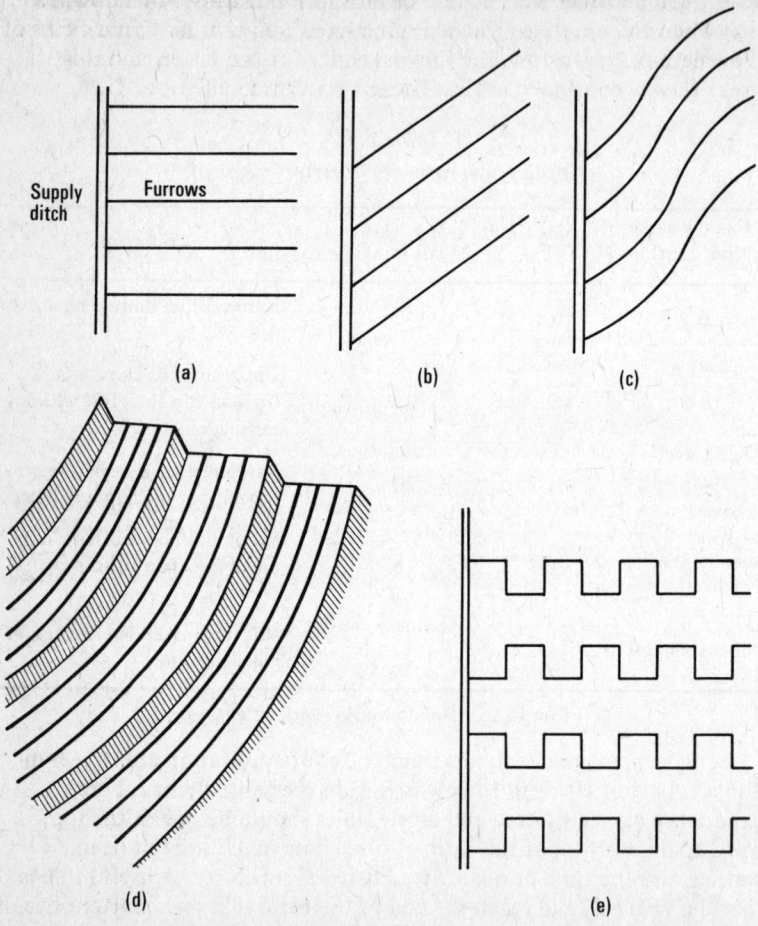

Figure 6.18.   Some furrow irrigation layouts: (a) furrows down the steepest slope; (b) furrows angled to reduce the gradient and the risk of erosion; (c) furrows following the contours; (d) furrows along the line of graded bench terraces; (e) furrows lengthened to increase the time for infiltration.

back up to the supply ditch at the top of the field, or by using it to supply another lower field. If re-use is not possible it must be drained off and discharged, otherwise the lower end of the furrows will be waterlogged.

Several possible furrow systems are shown in Figure 6.18. The most straightforward layout is to run the furrows straight down the steepest slope at right-angles to the supply ditch (Figure 6.18 (a)). If this will give too steep a gradient with the risk of erosion, the furrows may be angled if the field slopes uniformly (Figure 6.18 (b)), or they may follow contours (Figure 6.18 (c)). Land too steep for this may be bench-terraced with each terrace on grade so that furrows can be drawn along the terraces (Figure 6.18 (d)). If low infiltration makes it difficult to get enough water into the soil before it has flowed down the furrow, the time can be increased by making the furrows longer by making them in a zig-zag pattern (Figure 6.18 (e)). This is especially useful in orchards where large applications of water are required.

Length of furrow. The best length depends on slope, soil type, and the amount of water to be applied. If the furrow is too long the top end gets waterlogged, but if it is too short this means more supply ditches, and more interference with mechanization. If the gradient is very flat the water does not move quickly enough down the furrow, and if it is too steep there is the danger of erosion, so in both these cases the length should be less than in the case of the ideal slope which is about 0·3 per cent. Soils with high infiltration need short furrows because the water would sink in before it reached the end of a long furrow. The furrow length can be increased when larger amounts of water are to be applied. Table 6.18 shows some recommended sizes.

Grade of furrow. Furrow irrigation is most efficient on gentle uniform slopes where precise furrow grades can be obtained without too much levelling. The optimum furrow gradient is from 0·15 per cent to 0·3 per cent on heavy soils, and 0·3 per cent to 2·0 per cent on medium-textured soils. Furrows can be used on grades up to 5 or 6 per cent but it is difficult to avoid erosion in these cases, although it can be mini-mized by using small streams (Table 6.17).

Corrugations. Corrugation irrigation is really the same as furrow irri-gation but using small furrows and small water streams. Because the flow in each corrugation is small, it can be used on much steeper slopes than furrows, up to about 10 per cent. It is most often used for pasture and forage crops.

The corrugations are usually approximately semi-circular in cross-

Table 6.18
Length of furrows

(a) Metric units

Furrow slope (per cent)	Length of furrow (m)											
	Clay soils				Loamy soils				Sandy soils			
	Average depth of water applied (mm)											
	75	150	225	300	50	100	150	200	50	75	100	125
0·05	300	400	400	400	120	270	400	400	60	90	150	190
0·1	340	440	470	500	180	340	440	470	90	120	190	220
0·2	370	470	530	620	220	370	470	530	120	190	250	300
0·3	400	500	620	800	280	400	500	600	150	220	280	400
0·5	400	500	560	750	280	370	470	530	120	190	250	300
1·0	280	400	500	600	250	300	370	470	90	150	220	250
1·5	250	340	430	500	220	280	340	400	80	120	190	220
2·0	220	270	340	400	180	250	300	340	60	90	150	190

(b) English units

Furrow slope (per cent)	Length of furrow (ft)											
	Clay soils				Loamy soils				Sandy soils			
	Average depth of each irrigation (in)											
	3	6	9	12	2	4	6	8	2	3	4	5
0·05	1000	1300	1300	1300	400	900	1300	1300	200	300	500	600
0·1	1100	1400	1500	1600	600	1100	1400	1500	300	400	600	700
0·2	1200	1500	1700	2000	700	1200	1500	1700	400	600	800	1000
0·3	1300	1600	2000	2600	900	1300	1600	1900	500	700	900	1300
0·5	1300	1600	1800	2400	900	1200	1600	1700	400	600	800	1000
1·0	900	1300	1600	1900	800	1000	1300	1500	300	500	700	800
1·5	800	1100	1400	1600	700	900	1100	1300	250	400	600	700
2·0	700	1000	1100	1300	600	800	1000	1100	200	300	500	600

From F.A.O. Handbook, *Surface irrigation.*

Table 6.19
Length and spacing for corrugation irrigation

(a) Metric units

Slope (per cent)	Deep-rooted crops or shallow soils					
	Clay soils		Loam soils		Sandy soils	
	Length (m)	Spacing (m)	Length (m)	Spacing (m)	Length (m)	Spacing (m)
2	180	0·75	130	0·75	70	0·60
4	120	0·65	90	0·75	45	0·55
6	90	0·55	75	0·65	40	0·50
8	85	0·55	60	0·55	30	0·45
10	75	0·50	50	0·50		
	Shallow-rooted crops or deep soils					
2	120	0·60	90	0·60	45	0·45
4	85	0·55	60	0·55	30	0·45
6	70	0·55	50	0·50		
8	60	0·50	45	0·45		
10	55	0·45	40	0·45		

(b) English units

Slope (per cent)	Deep-rooted crops or shallow soils					
	Clay soils		Loam soils		Sandy soils	
	Length (ft)	Spacing (ft)	Length (ft)	Spacing (ft)	Length (ft)	Spacing (ft)
2	600	2·00	450	2·00	225	1·50
4	400	1·75	300	2·00	150	1·50
6	300	1·50	250	1·75	125	1·25
8	275	1·50	200	1·50	100	1·25
10	250	1·50	175	1·50	-	-
	Shallow-rooted crops or deep soils					
2	400	1·75	300	1·75	150	1·25
4	275	1·50	200	1·50	100	1·25
6	225	1·50	175	1·25	-	-
8	200	1·25	150	1·25	-	-
10	175	1·25	125	1·25	-	-

From F.A.O. Handbook, *Surface irrigation*.

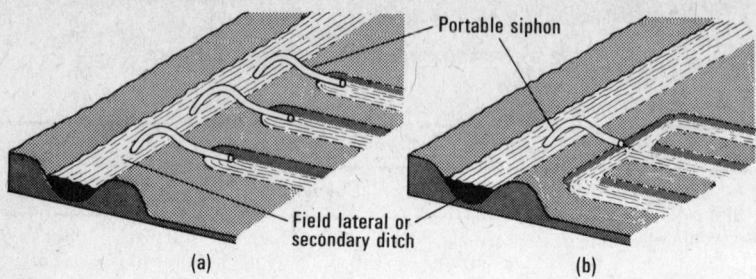

Figure 6.19.   Supplying water for corrugation irrigation; (a) siphon to each corrugation; (b) one outlet supplying several corrugations.

section, 75 - 100 mm (3 - 4 in) deep and 0·5 - 2 m (2 - 6 ft) apart. The length of run can be up to 300 m (1000 ft). Some lengths and spacings are shown in Table 6.19. Waterstreams are from 1 l/s to 2 l/s (10 - 20 gall/min) per corrugation. One technique for applying the water is to start with a full flow and then cut back to a quarter (like furrow irrigation). The other method is to treat it like border strip and cut off completely when the water has travelled four-fifths of the distance. The water may be applied through individual syphons to each corrugation, or one outlet from the supply ditch can feed a group of several corrugations (see Figure 6.19).

7. Drainage

7.1. Drainage needs and benefits

7.1.1. The need for drainage

It is convenient to divide drainage into two types. *Land drainage* is large-scale drainage where the object is to drain surplus water from a large area by such means as improving the flow of the streams and rivers, excavating large open drains, erecting dykes and levees, and pumping. Schemes of this nature will be associated with large areas of low-lying land, frequently in coastal areas. Such schemes are major civil engineering works, and will not be discussed here.

The kind of drainage which concerns agricultural engineers is *field drainage*. This is the removal from agricultural land of surplus water which might otherwise restrict crop growth. Some of the adverse effects of excess water are to reduce the aeration of the soil, to reduce soil temperature, to inhibit root growth, or to reduce the volume of soil available for root growth. The problem may be localized, for example, wet patches in a field, or may be over a large region. It may occur temporarily for a period of days or weeks, or may be a permanent condition. Crops vary greatly in their tolerance of wet conditions, both as to the amount of excess water they can stand and the duration of the flooding, but almost all commercially grown crops are adversely affected to some extent by inadequate drainage.

The surplus accumulates because incoming water, as rain or surface flow, cannot naturally drain away fast enough, and the practice of field drainage is directed towards accelerating or increasing the natural out-flow, either on the surface by means of open drains or ditches, or below the ground by a system of closed underdrains. If the primary object is to avoid surface waterlogging then surface drainage is indicated, but if a permanent lowering of the water table is desired, then a system of underdrains is more often used.

In European countries underdrainage has been practised for centuries, and a high proportion of the land has been drained. This is partly because of the relatively low permeability of most of the soils, and partly because intensive high-capital farming has been practised for a long time. In developing countries in the tropics and sub-tropics drainage is mainly surface drainage, but the amount of subsurface drainage can be expected to increase as new techniques and materials are able to lower the cost.

Two special cases should lead to more emphasis on field drainage in developing countries. The need for drainage to be incorporated into irrigation schemes is now generally accepted, although this was not always so in the past. The worldwide expansion of irrigation will therefore lead to more drainage. The second factor, also often associated with irrigation, is that when there are problems of salinity, a good drainage system is nearly always required for any method of control or reclamation. Many millions of hectares have gone out of production in Iraq and Iran, and India and Pakistan, because irrigation has raised the water table and brought up salts. After the soil has been brought to field capacity a further application of 25 mm of irrigation water can raise the water table by 150 - 250 mm (6 - 10 in), depending on soil type. It therefore does not take much over-watering to bring the water table up to the surface. In addition to over-irrigation, drainage problems on irrigation schemes can arise from unexpected rain after irrigation or from seepage losses from storage reservoirs or canals.

As with most agricultural problems, the first step towards solving a drainage problem is to identify the cause. Sometimes what appears to be a major drainage problem can be solved by a simple remedy once the cause is recognized. An example of this is waterlogging which is caused by water coming in from outside the wet area. This can often be remedied by a single drain which intercepts the flow. An open ditch will intercept any surface flow, and also subsurface flow if it is dug deep enough or down to an impervious layer as in Figure 7.1. On rolling land it is nearly always worth trying the effect of an interception drain before putting in a full drainage scheme.

Drainage coefficient. When designing a drainage scheme, the amount of water to be removed has to be known. Most growing crops suffer little damage if they are waterlogged for short periods so it would be un-economic to put in schemes capable of draining away the excess water in a few hours. It is usual to specify the amount which is to be drained in 24 hours, and this is the definition of drainage coefficient. It is expressed as the depth of water, over the drained area, which is to be removed in 24 hours.

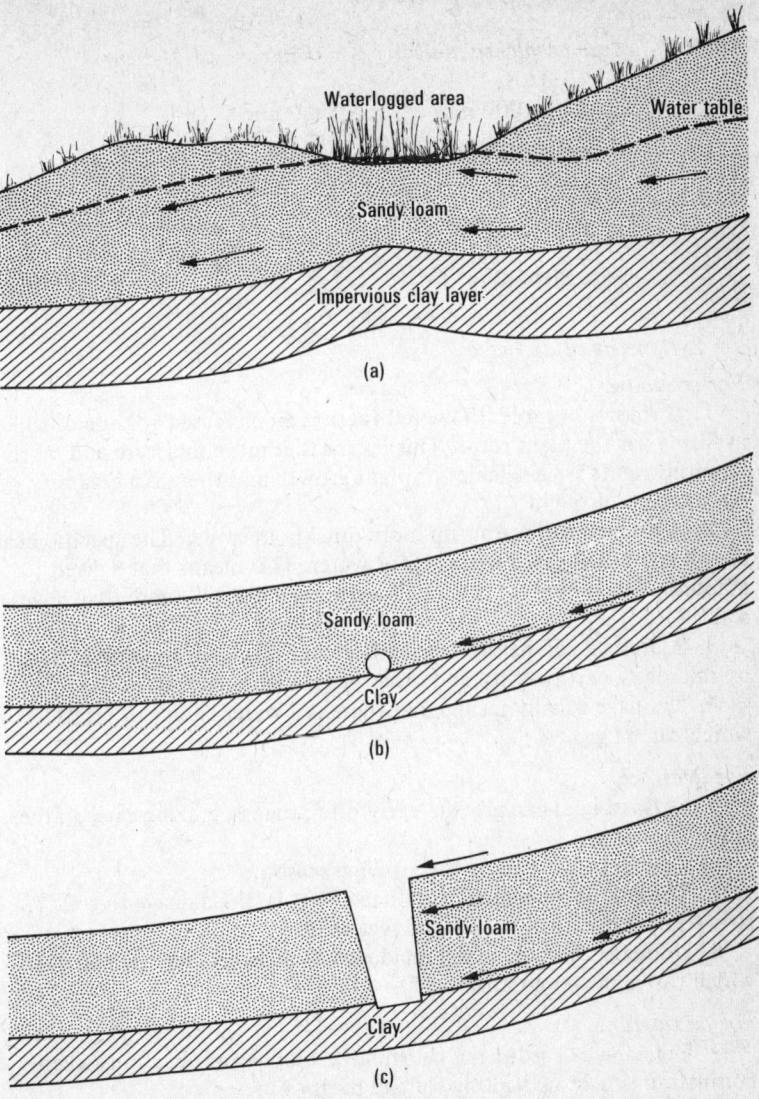

Figure 7.1. The use of interception drains: (a) waterlogging caused by subsurface inflow; (b) subsurface drain intercepts subsurface flow; (c) open drain intercepts both surface and subsurface flow.

Typical figures are:

Mean annual rainfall (MAR)	Drainage coefficient
less than 1000 mm	10 mm
more than 1000 mm	$10 \times \dfrac{MAR}{1000} = \dfrac{MAR}{100}$ mm

or in English units:

less than 40 in	$\frac{3}{8}$ in
more than 40 in	$\frac{3}{8} \times \dfrac{MAR}{40}$

7.1.2. Benefits of drainage

For cropping.

1. If the water table is lowered there is an increased volume of soil available for the plant roots. This means that more moisture and more plant nutrients are available for plant growth and there is a greater resistance to drought.

2. A drained soil warms up more quickly in spring. The specific heat of soil is 0·2 compared with 1·0 for water. This means that a given amount of incoming sunshine will warm up a dry soil more than a wet soil.

3. A drained soil has better aeration and this leads to more activity by microbes, earthworms, and other beneficial organisms in the soil.

4. Drainage usually means that there is a wider choice of crops which can be grown.

For livestock.

5. Drained land can give an 'early bite', that is, grazing early in the growing season.

6. Drainage may extend the growing season.

7. The risk of 'poaching' is reduced, that is, the damage to soil structure by livestock walking on wet soils.

8. Drainage can reduce the incidence of diseases such as liver-fluke which thrive in wet conditions.

For cultivation.

9. Drainage can reduce or shorten the number of occasions when cultivation is held up waiting for soil to dry out.

10. There is likely to be less damage to soil structure by working the soil in too wet a condition.

11. Drainage can make it possible to harvest crops earlier and in dry conditions which will lead to obtaining a better quality.

7.2. Surface drainage

The object of surface drainage is to increase the removal of surface water either by eliminating low spots where the water might tend to accumulate, or by excavating drains and ditches to speed up the water movement. In many cases a small increase in the efficiency of the surface drainage is sufficient to halt or reverse an undesirable trend.

An interesting example of this can be quoted from personal experience in Africa. The soil over a large region was fertile well-drained loam, almost flat—very good farm land. The natural vegetation was tree savanah, typical of a 500 - 700 mm (20 - 30 in) rainfall concentrated in a 5-month season. Large areas of this country were cleared of their natural vegetation and used for annual crops such as maize, cotton, and sorghum. As a result the moisture consumption was slightly less than under the natural tree growth, not by very much, perhaps only 25 millimetres (1 in) or so during the year. However the difference, that is the moisture not used by the annual crops, went down to the ground water, and the water table started to rise. Records of wells showed that under natural vegetation the water table at its highest point during the wet season rose to about 5 m from the surface. Under annual crops it started to rise by an average of 100 mm (4 in) per year. After 40 years of cropping the stage was reached where the raised water table was seriously interfering with cropping. However, the difference between the original water regime and that under annual cropping was small, and a simple system of surface drainage was sufficient to control the problem by ensuring that an extra 20 mm (1 in) or so of rain became surface run-off.

7.2.1. Land forming

The simplest way of improving the surface drainage is just to level the soil to an even grade so that minor irregularities are evened out. The techniques for working out how to balance the soil cut from the high places and the fill in the low spots were described in Section 2.2.6 on levelling land for surface irrigation.

The drainage is improved more by *bedding*, that is, working the land into a regular pattern of raised beds. This has two effects. The soil in the raised beds drains better and quicker, and the ditches get the water away better. There is a wide range of different styles of bedding for different purposes. Old grassland in England nearly always has high, narrow beds, perhaps only a metre apart and half a metre between the top of the bed and the bottom of the furrow (Figure 7.2 (a)). The

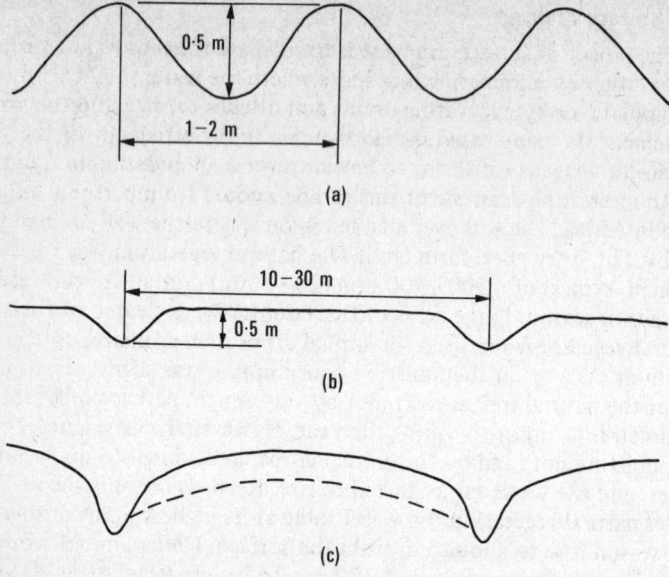

Figure 7.2.   Some forms of bedding: (a) old English grassland; (b) ridge-and-furrow method; (c) camber-bed method.

bedding is straight up and down the slope, and used even on steep slopes.

A variation more suitable for mechanized farming on arable land is called *ridge and furrow*. It is used most on gentle slopes up to about 2 per cent. The ridges are wider, anything from 10 m to 30 m (30 - 100 ft), with the centre about 0·5 m above the furrow. The ridges and furrows are set out diagonally to the steepest slope, so that there is a fall down the furrow and also in the drainage ditch into which the furrows discharge (Figure 7.2 (b)). The ridges are formed by repeated ploughing inwards from the furrows, perhaps followed by the use of a light grader to get a smooth regular shape.

Heavy soils with poor internal drainage may need something more than just increasing the rate of surface drainage, and in this case the answer may be the kind of ridge and furrow called *camber-bed cultivation* (Figure 7.2 (c)). The ridges are built up higher, with a cambered surface sloping down to the open drain on either side. Cultivating to a constant depth below the cambered surface means that the top of the undisturbed subsoil also slopes towards the furrows and this encourages the movement of water to the drains.

In ridge and furrow systems the direction of ploughing and cultivation is along the ridge for crops which are grown on the flat, but crops grown on ridges such as tobacco, pineapples, and maize are sometimes planted at right-angles across the main ridges. This can make cultivation more awkward unless the furrows are small and easily crossed by tractors and implements.

7.2.2. Open ditches

A wide variety of open ditches are used for draining different kinds of land. For permanent pastures the drains are small, shallow, and closely spaced. For arable lands the ditches are larger and more widely spaced.

For poorly drained fields with gentle slope, up to about 2 per cent, parallel field drains may be sufficient and simpler than a ridge-and-furrow system. The drains will have a wide,shallow cross-section to minimize the risk of erosion and to allow easy crossing. The grade will be dictated by the slope of the land, but should be between 0·1 per cent and 1·0 per cent, with 0·5 per cent as the optimum. The minimum depth is about 250 mm (10 in) and the spacing can be up to 200 m.

For slightly steeper slopes, from 2 per cent to 4 per cent, the system of cross-slope ditches is often used, partly for surface drainage and partly to control soil erosion. Open ditches are run on a gentle gradient (preferably 0·5 per cent) across the slope, just like channel terraces (Section 8.3.2) without the bank on the lower side. The drains are wide, shallow, and dish-shaped—about 5 or 6 m (15 - 20 ft) wide at the top and 150 - 200 mm (6 - 9 in) deep.

7.2.3. Machinery for ditching and draining

Digging ditches and drains by hand may be the cheapest or the most expensive method, depending on the cost of labour. Apart from the question of cost it may be appropriate to use machinery for other reasons such as working more quickly, or in harder conditions than would be possible with hand labour.

The two main jobs connected with surface drains are excavating the open drains and cleaning and maintaining them. Large open drains such as cross-slope ditches are excavated by earth-moving machinery, such as bulldozers, scrapers, and powered graders, or by farm implements, such as wheel tractors drawing ploughs, disc-harrows, or light graders.

Two special kinds of drainage plough deserve mention. One is like a very large mouldboard plough which allows large open drains up to

1·5 m deep to be ploughed out in a single operation. An extension of the mouldboard pushes the soil back from the edge of the trench. A very large pull is required, and this is provided by several crawler tractors connected in series.

The other kind of drainage plough is smaller and like a ridger, or two mouldboard ploughs back to back. This cuts the drain to the right shape in one pass and piles the soil on either side. The plough may incorporate a special hitch which gives a smooth even grade to the drain, even if the surface is undulating. This kind of plough can be very useful for the first stage of reclaiming derelict swamp or bog as discussed in Section 2.3.1.

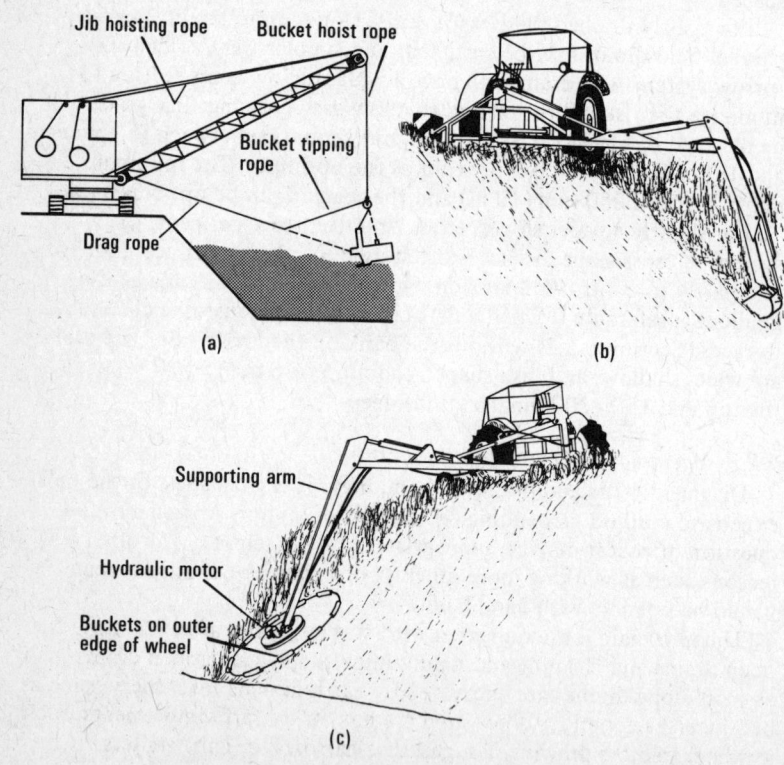

Figure 7.3.   Machinery for maintaining open drains: (a) drag-line bucket; (b) back-acter; (c) rotary ditch cleaner.

The maintenance of open drains through arable land is no problem as they can be kept free from weeds, and the shape of the drain maintained as part of the cultivation operations. But other open ditches do need a special cleaning operation, and many kinds of machines have been developed for this purpose. Most of them operate from one side of the drain. A few machines straddle the drain, and this does allow a symetrical line of work, but getting the machine on and off the drain needs a bridge or closed section, and also a lot of ditches have one side very close to a fence or hedge.

If the main purpose of the cleaning operation is to deepen the ditch or remove silt from the bottom, the dragline bucket is most used (Figure 7.3 (a)). The machines used for trimming the banks or removing vegetation are usually one of the many varieties of backacter or back-hoe. Some machines are controlled by cables like the dragline; others have a scoop on a movable jointed arm controlled by a number of hydraulic rams. Some of the larger machines of this type are purpose-built and do only this work; others are mounted on the rear of a farm tractor and operate from its hydraulic system. (Figure 7.3 (b)). Another principle used for bank-cleaning is a large-diameter wheel with buckets fixed round the outer edge and driven by a hydraulic motor. Hydraulic rams control the positioning and angle of the wheel (Figure 7.3 (c)).

7.3. Subsurface drainage

Although the basic objective of both surface and subsurface drainage is to provide a drier soil for plant growth, the way this is achieved is quite different. Surface drainage aims at increasing the surface run-off and so reducing the amount of water going into storage in the soil. Subsurface drainage aims at increasing the rate at which water will drain from the soil, and so lowering the water table, thus increasing the depth of drier soil above the water table. The effect of subsurface drainage on crop growth is illustrated in Figure 7.4.

The movement of water to subsurface drains is shown in Figure 7.5. Drain pipes are laid in a trench and then covered over, but the effect is as though the open drain were still there, that is, water moves towards the drain and flows away down it. The rate at which water can move through the soil is called its hydraulic conductivity, and this has a big effect on the design of subsurface drainage.

There are many methods for the calculation of hydraulic conductivity from tests carried out in the field or on samples in the laboratory. The problem is that, when sophisticated equipment and complicated

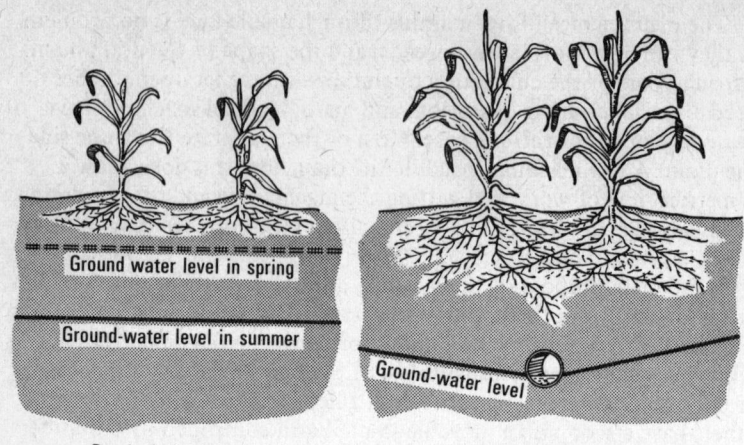

Poorly drained land   Tile-drained land

Figure 7.4.   The effect of drainage on crop growth.

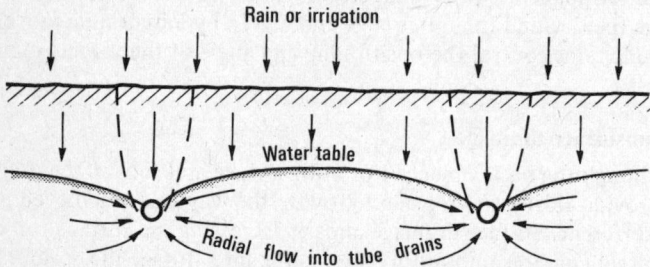

Figure 7.5.   The movement of water to subsurface drains.

equations are used to obtain precise values, these usually vary so much from one point in the field to another that there seems little point in seeking great accuracy in individual measurements.

The following version of the auger-hole method makes many assumptions and simplifications but will give results good enough for practical drainage design. (The list of *Further reading* (p. 219) gives more detailed accounts of the method.)

The limitations are that the soil must be fairly uniform, without an impermeable layer near the depth where the drains will be laid, and having a slow to medium permeability (that is free-draining sandy soils

and very slow-draining clays need special consideration). There must be
a water table close to the surface.

A hole is made with a soil auger 80 mm (3 in) in diameter and
extending between 600 mm and 700 mm (24 - 28 in) below the water
table. The 'Dutch auger' shown in Figure 1.2 (p. 4) is very suitable.
The hole is pumped or bailed out and allowed to refill several times,
thus flushing out any fine particles which may have been forced into
the side walls during the process of boring. To carry out the test, the
water level is lowered by pumping or bailing and then the time is
measured that the water takes to rise a fixed distance. The simplest
method to locate the water level is a small float fastened to the end of a
metal tape-measure, but electrical devices or air-bubbling methods are
also used (Figure 7.6). The measurement of the water rise must be done

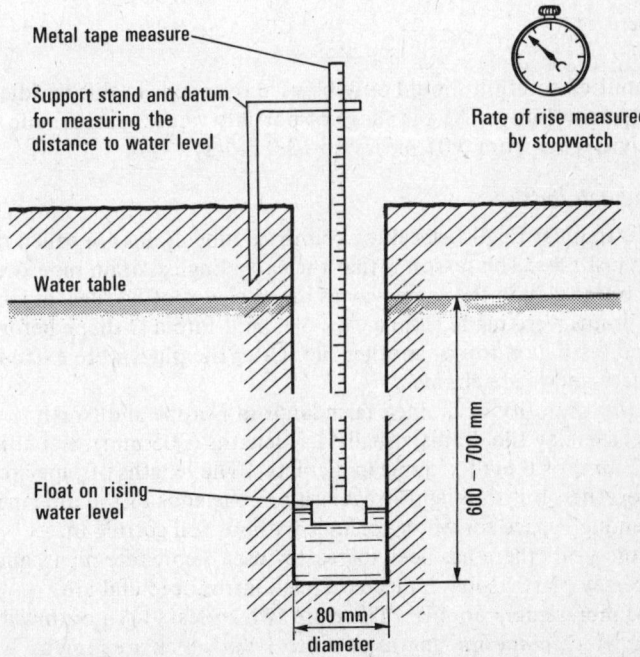

Figure 7.6.   The auger-hole method for determining hydraulic conductivity. (After van Beers, Bull. No. 1, Int. Inst. for Land Reclamation
and Improvement, Wageningen, The Netherlands.)

quickly, before the water has risen to more than a quarter of the distance pumped out.

Hydraulic conductivity varies greatly, and so will the rate of rise of the water level in the hole. The rate of rise is required in units of millimetres per second, and convenient distances to measure and possible times are shown in the table below. Finally, the rate of rise is multiplied by 0·8 to give hydraulic conductivity in metres per day.

Soil type	Suggested range of measurement (mm)	Example time (secs)	Rate of rise (mm/s)	Hydraulic conductivity K (m/day)
Moderately permeable	100	100 to 200	1·0 to 0·5	0·8 to 0·4
	50	500	0·1	0·08
Slowly permeable	25	500	0·05	0·04

This simplified solution should only be used over soils in this middle range of permeability. The full range of possible values of hydraulic conductivity goes from 0·01 m/day to 10·0 m/day.

7.3.1. Tube drains

Materials. In most English-speaking countries tube drains are called *tile drains* or just *tiles*. The reason is that the early English drain pipes were made of burnt clay in the same way as burnt-clay roofing tiles, in fact the first drains were made from a clay tile, bent into a U-shape before firing, and resting on top of another tile. Later the pipes were extruded in one piece and made circular.

The great majority of drained farm lands in Europe and North America have clay tile drains, usually 1 ft lengths (305 mm), and either 4 in (102 mm) or 6 in (152 mm) in diameter. The lengths of pipe are butted together, but the slight irregularity of the ends of the pipe means there is enough space for water to enter without soil getting in.

Over the years there has been increasing use of concrete pipes, and more recently plastic tube drains have been introduced and are becoming increasingly popular. There are two styles: (1) smooth-walled rigid pipe which comes in lengths of up to 10 m which are joined together in the field as they are laid, and (2) corrugated pipe which has sufficient flexibility to be rolled in coils or 2 or 3 m diameter, and which allows the pipe to be supplied in lengths up to 200 m. Small slits at frequent intervals along the pipe allow the entry of water in the same way as the joints in tile drains.

There are two main advantages of plastic pipe.

1. It is very much lighter in weight than either clay or concrete pipes. This lowers the transport costs which are a big item for clay and concrete pipes, and also makes it easier to get the pipes onto the site which needs drainage and so is quite likely to be unsuitable for heavily laden vehicles.

2. A smaller-diameter pipe can be used to carry a given flow of water, partly because there are no joins as in the case of tile drains and partly because the inside surface is smoother, that is, it has a lower roughness coefficient (0·0094 compared with 0·0108 for clay and concrete).

The smallest size of clay or concrete pipe in regular use is 100 mm (4 in) diameter, but plastic pipes of half this size (50 mm diameter) are satisfactory.

The gaps between tile drains, or the slits in the plastic tubes, allow water to enter the drains, but it is also necessary for water to be able to move freely up to the pipes. To help this movement it is common to place round the pipe some free-draining material called a *permeable fill*. This may be a graded gravel or coarse sand. Plastic materials such as expanded polystyrene are, like plastic tubing, increasingly popular because of their light weight. Another significant saving which results from using plastic pipes is that less permeable fill is needed in the smaller trench.

Laying plastic pipes. The smooth-walled rigid plastic pipe is laid in a similar manner to tile drains. That is, a trench is excavated, the pipe is laid in the trench, permeable fill is added, and then the trench is filled. The smaller size means a big reduction in the excavation cost and the amount of permeable fill.

The flexible corrugated pipe has made possible the development of machines which lay the pipe without excavating an open trench. A powerful tractor draws behind it a tool-bar carrying an implement like a subsoil ripper or a mole plough (Figure 7.7). At the bottom of the shank is a pointed cylinder which forms a round hole as it is drawn through the soil. The vertical shank is hollow and wide enough for the pipe to be fed down through it as the tractor moves across the field. The operation has to start at an open ditch or a hole dug for the purpose. The corrugated pipe is carried in large coils on the tractor, and the operation is only interrupted to join on a new coil, usually every

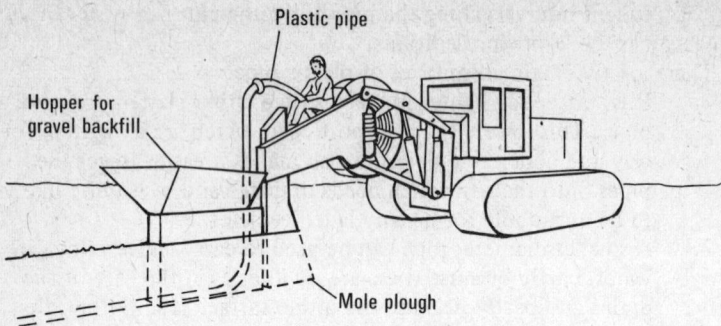

Figure 7.7. Trenchless tube drainage. (From Withers and Vipond, *Irrigation: design and practice*, Batsford, London.)

200 m (650 ft). Another advantage of this method is that the permeable fill can also be dropped down the hollow shank so that the plastic pipe has over and around it a layer of porous material through which the water can pass easily to the pipe. This ensures that the pipe will always be running at full capacity.

7.3.2. Subsurface drainage design

The drainage coefficient and the area together determine the required rate of flow from the field, that is, the outlet discharge. Figure 7.8 shows how this can be used to determine alternative combinations of size and gradient. To take a particular example, a field of 20 ha with a drainage coefficient of 10 mm in 24 hours will have a discharge of 23 l/s. The alternative pipes are those on the horizontal line of 23 l/s and are

150 mm pipe at grade of 2·5 m in 100 m or $\frac{1}{40}$

200 mm pipe at grade of 0·5 m in 100 m or $\frac{1}{200}$

250 mm pipe at grade of 0·1 m in 100 m or $\frac{1}{1000}$.

Comparable English unit equivalents are given in Table 7.1.

Design variables. The effect of the main variables in a subsurface drainage scheme are shown in Figure 7.9. The kind of soil affects the hydraulic conductivity, which determines the shape of the draw-down curve towards the drains. A sandy soil will have a high hydraulic conductivity, that is, the water moves easily so the draw-down curve is flat, compared with a clay soil (Figure 7.9 (a)). The second diagram

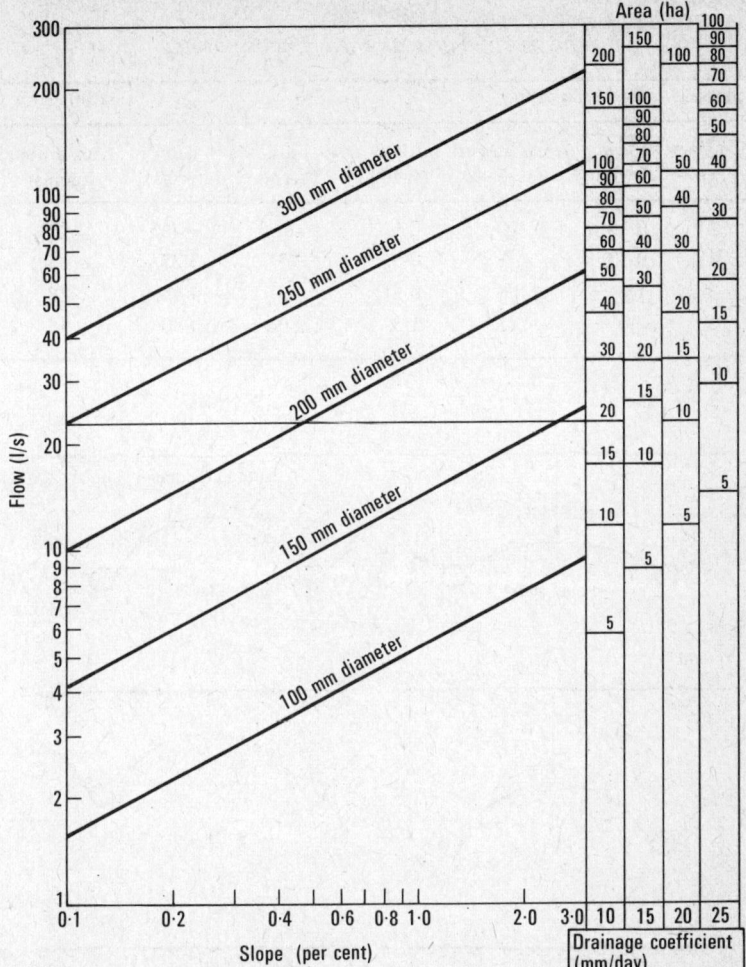

Figure 7.8.   The discharge from tile drains.
Example. 20 ha at 10 mm/day needs 150 mm pipe at 2·5 per cent or
200 mm pipe at 0·5 per cent or 250 mm pipe at 0·1 per cent.

(Figure 7.9 (b)) shows that, other factors being equal, the greater the
depth of the drains the greater the depth of drained soil. Ensuring an
adequate depth for root development is a primary objective. The

Table 7.1
Pipe size and area drained (English units)

Slope	1 : 100		1 : 250		1 : 500	
Pipe size (in)	Flow (ft³/s)	Area drained (acres)	Flow (ft³/s)	Area drained (acres)	Flow (ft³/s)	Area drained (acres)
4	0·15	10	0·1	6	0·065	4
6	0·55	35	0·35	23	0·25	15
8	1·0	65	0·6	40	0·45	28
10	2·0	125	1·3	80	0·80	55

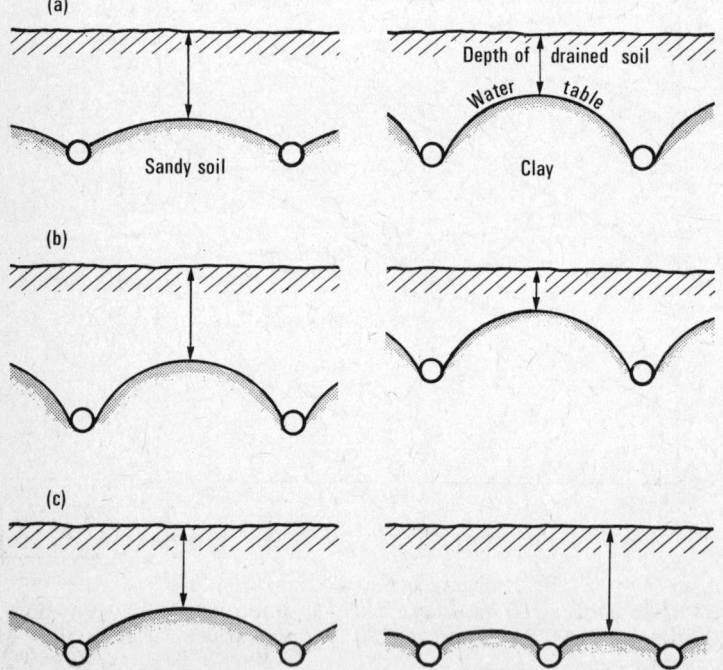

Figure 7.9.   The effect of the variables in the design of tube drainage. (a) Effect of soil type. (b) Effect of depth of tube drains. (c) Effect of spacing of tube drains.

Table 7.2
Depth and spacing of tube drains (metric units)

Soil type	Depth		Spacing	
	(m)	(ft)	(m)	(ft)
Sand	0·6	2	up to 60	up to 180
Sandy loam	0·8 - 1·0	2·5 - 3·0	up to 60	up to 180
Silt loam	0·8 - 1·8	2·5 - 5·5	20 - 80	60 - 240
Clay loam	0·6 - 0·8	1·8 - 2·5	15 - 100	45 - 300
Peat	1·2 - 1·5	3·6 - 4·5	40 - 60	120 - 180

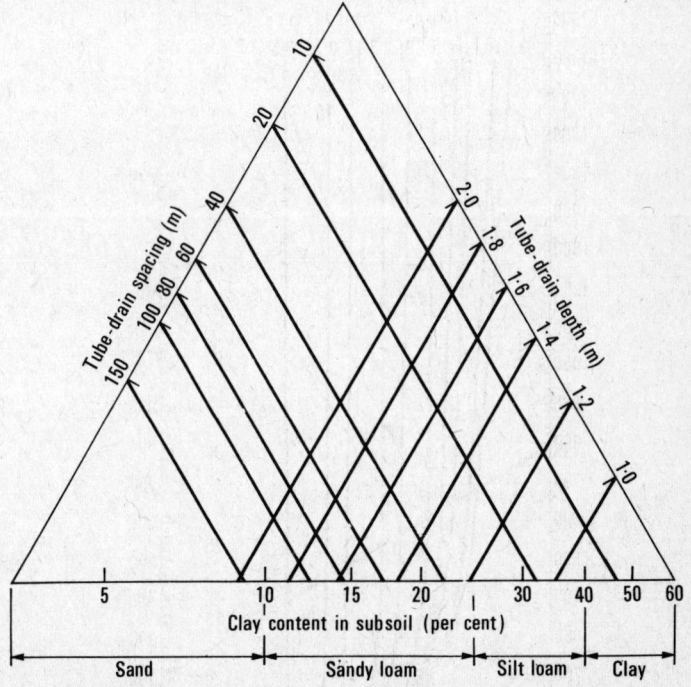

Figure 7.10.   An American design chart for depth and spacing of tube drains. The spacing is wider and the depth greater than the European practice shown in Table 7.2. (Minnesota Agricultural Experimental Station, Tech. Bull. 101.)

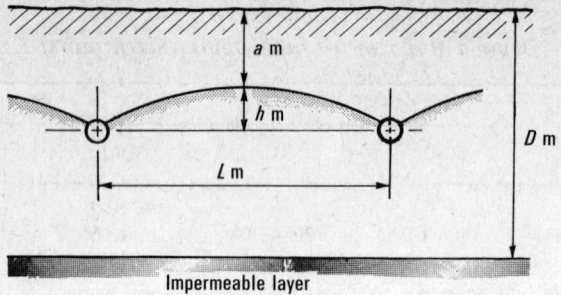

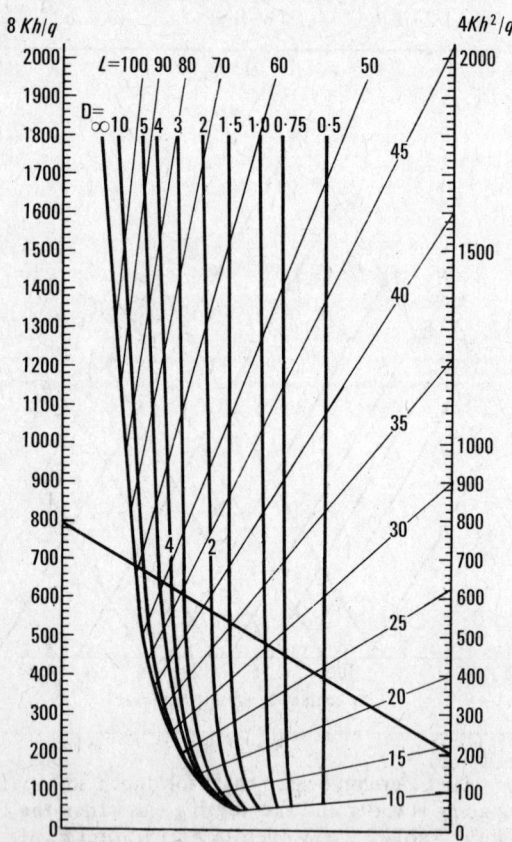

Figure 7.11.   A design chart from the Netherlands for depth and spacing of tube drains. (After Ernst, Bull. No. 8, Int. Inst. for Land Reclamation and Improvement, Wageningen, The Netherlands.)

spacing of the drains also affects the minimum depth of drained soil, which will be greater when the drains are closer together (Figure 7.9 (c)).

Naturally these three variables interact upon each other, and several procedures are available for choosing suitable designs. Table 7.2 and Figure 7.10 both give a range of depth and spacing combinations for different soil types.

More rigorous solutions based on measurements of the hydraulic conductivity have been developed in the Netherlands, for example, the *Hooghoudt method*. The procedure is as follows.

1. Measure the hydraulic conductivity K (m/day).
2. Select a value for the required draw-down h (m) and the required depth a (m) of drained soil. The depth of the drains will thus be $(a + h)$ (Figure 7.12).
3. Select a value for the required drainage rate q (m/day).
4. Calculate $8Kh/q$ and locate this value on the left side of the chart in Figure 7.11 and calculate $4Kh^2/q$ and locate this value on the right side.

The line joining these two points gives alternative spacings, depending upon whether there is an impermeable layer, and its depth D (m). If there is no impermeable layer the depth is taken as infinity.

Example
$a = 0.75$ m, $h = 0.5$ m, $q = 0.005$ m/day, and $K = 1.0$ m/day. The drains will be $a + h = 1.25$ m deep.
$$\frac{8Kh}{q} = 800, \qquad \frac{4Kh^2}{q} = 200.$$
If there is an impermeable layer at 3 m, that is, $D = 3$, the spacing should be 45 m. If there is no impermeable layer $D = \infty$ and the spacing should be 60 m.

Some practical details: Backfilling. To reduce the problem of silt getting into the pipes through the gaps something can be placed on top of the pipe. In the old days this was usually sods of turf placed grass-side down on top of the pipe, but nowadays it is more likely to be a strip of plastic film, glass fibre, or waterproof paper. The term *blinding* is given to this first covering of the pipe before the main filling and it may also include dropping a small amount of soil on each side of the pipe to prevent it being dislodged during the main backfilling which is often done by machine. Some cross-sections showing completed drains are shown in Figure 7.12.

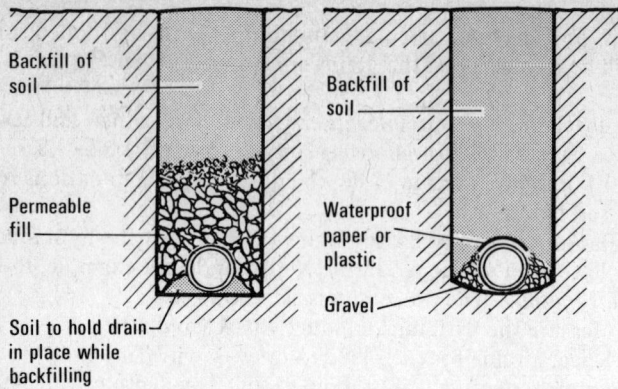

Figure 7.12.   Examples of tube-drain installation.

Blockages of subsurface drains are difficult to deal with because the exact trouble spot is hard to find, and it is best to take care to minimize the risk of blockages occurring. The main causes are:

(1)   blocked outlets—and the remedy is to keep outlets clean and to fit guards to prevent animals such as rats from getting in;

(2)   silting up—and the remedy is careful grading, blinding on top of the pipes, and allowing only small gaps between tiles;

(3)   blockages from roots of trees or shrubs—and the remedy is to remove these from near drainage lines.

7.3.3. Mole drains

Cheap unlined underdrains, called mole drains, are effective in clay soils which will retain a moulded shape after a mole plough (Figure (7.13) is drawn through the soil. The vertical shank and the mole shatter

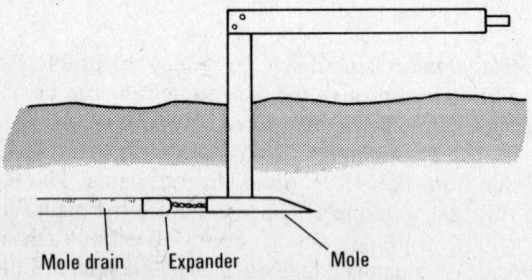

Figure 7.13.   A mole plough.

and crack the soil to increase the downward movement of water, and the expander shapes a circular smooth-walled tunnel which carries the water. By mounting the plough at the end of a long beam the depth becomes self-adjusting and evens out some of the surface irregularities.

For efficient operation the soil must be suitable and the drains must be drawn when the soil is at the right moisture. Mole drains are shallower and closer than tile drains, usually from 300 mm to 750 mm (1 - 2·5 ft) deep and from 1·5 m to 10 m (5 - 35 ft) apart. Gradients may be between 1 in 50 and 1 in 500. Diameters up to 200 mm (8 in) have been used, but 60 - 90 mm (2·5 - 3·5 in) are most common. A working life of 10 - 15 years is normal in good conditions.

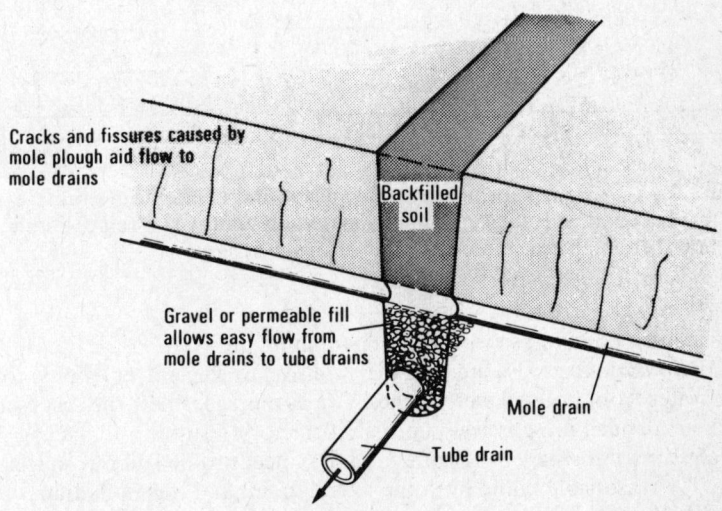

Figure 7.14.   The junction of tube and mole drains.

Combined systems. The weakness of mole drains is their outlets, which are subject to collapse and vulnerable to interference. By allowing a network of mole drains to discharge into tiled main drains the advantages of each can be obtained. The tile mains are installed first at greater depth then shallower mole drains are drawn at an angle to the tile drains. At intersections the trench carrying the tile drain is filled with gravel (Figure 7.14).

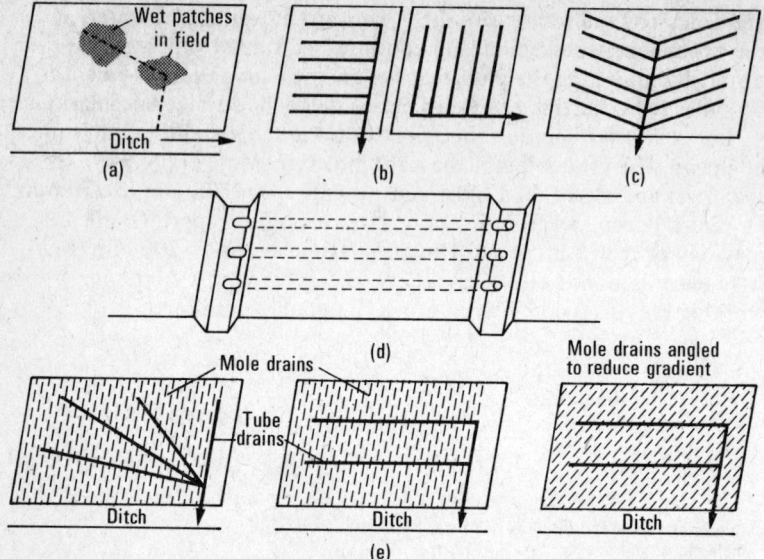

Figure 7.15.   Some subsurface drainage layouts: (a) random drainage; (b) regular parallel drainage; (c) herring-bone layout; (d) ditch-to-ditch method; (e) combined mole and tube drainage schemes.

7.3.4. Drainage layouts

The most common schemes are shown in Figure 7.15.

Random drainage (Figure 7.15 (a)) consists of single lines of pipe to take water from isolated wet patches. The advantage is that the scheme can be expanded progressively until a satisfactory result is obtained.

Regular parallel layout (Figure 7.15 (b)) needs to be laid out so that there is a reasonable fall in both the lateral drains and the main drain into which they run.

Herring-bone layout (Figure 7.15 (c)) gives a good fall in the mains and some fall in the side drains. It requires a lot of junctions between main and side drains.

Ditch-to-ditch layout (Figure 7.15 (d)) may be suitable for free-draining soils with little gradient, where deep, open drains are required for water-table control.

In *combined mole and tile-drain layouts* (Figure 7.15 (e)) the purpose of the mole drains is to lead the water to tile drains which take it away to the ditch or discharge point.

8.   Soil conservation

8.1. Principles of erosion control

The amount of soil erosion caused when rain falls on farm land depends on three things: the nature of the rain, the kind of soil, and the way the land is being farmed. The ability of rain to cause erosion is called the *erosivity*. Naturally the greater the amount of rain the greater is its power to damage the soil, but the kind of rainfall also makes a difference. In fact the erosivity depends upon the kinetic energy of the rain (kinetic energy is the energy possessed by any moving body—mathematically, kinetic energy $=\frac{1}{2} \times$ mass $\times$ velocity^2). Erosivity is highest for the kind of rain common in the tropics and sub-tropics, that is, thunderstorm rain with large drops and high intensities. The erosivity of rain can be calculated, but we cannot do anything about changing it to reduce erosion.

Neither can we do much about the kind of soil. Some soils are more easily eroded because of their chemical and physical properties. In general, light or sandy soils are more easily eroded than heavy or clay soils. The extent to which a soil is vulnerable to erosion is defined as its *erodibility*. This too can be measured, but little can be done to change the basic texture of a soil, so we cannot alter erosion this way.

The third factor affecting erosion is the way the land is managed, and only here can we influence and control it. However, the difference in soil loss resulting from different management is tremendous. The same land can suffer catastrophic erosion under one form of land-use, and none at all under another.

There are two ways in which our management of the land affects erosion. The first is to decide what kind of farming is best, or what rotation to follow, or how a particular crop should be grown—these are all included in what we call *biological erosion control*. The other aspect is the physical control of soil and water movement by the use of drains, banks, terraces, and other earth-moving methods, and this is called *mechanical erosion control*. Effective soil conservation requires an understanding of the different role of each method, and how to combine them both for effective control of erosion.

8.2. Biological control

8.2.1. Land-use

The most fundamental principle of soil conservation is that it can only be effective when the land-use is ecologically sound. Trying to bolster up an unsound form of land-use by conservation measures is like giving aspirin to a man suffering from a dislocated shoulder. It may provide temporary relief, but the only satisfactory long-term solution is to put right the basic cause of the problem. For example, when cereal crops are grown on steep mountain sides it may be possible to reduce the erosion by building terraces, but cheap terraces will not last, and long-lasting terraces will cost more than the crop is worth. However, good soil conservation does not conflict with profitable agriculture. If the land is too steep for cash crops these will give a poor return, but a suitable land use such as a well-managed pasture will be more profitable. The methods for classifying land according to its best potential use were discussed in Chapter 1.

8.2.2. Crop management

This is by far the most important aspect of soil conservation simply because it has the greatest effect. By building terraces on arable land we

Figure 8.1.   This diagram is copied from a high-speed photograph of the moment when a raindrop lands on wet soil. The effect is like that of an explosive bomb.

may reduce the erosion by half, and this is well worthwhile. But by changing the crop management we can reduce the erosion to a tenth or a twentieth of what it would be if the same crop were grown a different way.

To understand the reason for this we must look at the actual process of soil erosion. The old-fashioned explanation of soil erosion was that it is caused by water flowing over the surface and washing the soil away. In fact this is only part of the process, and not the most important part. Erosion starts when falling raindrops splash onto the soil. This has been studied by the use of high-speed photography, which shows that each raindrop which strikes the soil has an effect like a bomb bursting (Figure 8.1). Several results follow.

First, the splashed particles move farther downhill than up, so there is a net movement downhill (Figure 8.2). An example of this process is that if a pile of building sand is left in the rain it gets flattened, although there is no surface run-off at all.

Figure 8.2.   Soil is splashed farther down the slope than up the slope, so there is a net movement down the slope.

Second, the splash breaks up the soil aggregates into smaller particles which are more easily washed away.

Third, these finer particles fill the spaces between the aggregates, and so reduce the infiltration and increase the run-off. Many research workers have shown that by preventing splash the erosion from bare soil can be dramatically reduced. For example, in an experiment in Africa the 10-year average soil loss from bare soil was 126·6 tonne/ha (50·6 ton/acre) but soil completely protected from splash by fine wire gauze lost only 0·9 tonne/ha (0·36 ton/acre).

The same effect can be achieved on a field scale by using growing crops to provide the protective cover. One very effective method for row crops is to increase the plant population (that is, the crop density), so that more ground is covered. For example, maize grown at 25 000 plants per hectare (0·4 m apart in 1 m rows) lost in 1 year 12·3 tonne/ha of soil, but when the plant population was increased to 37000 per hectare (0·27 m apart in 1 m rows) the soil loss was reduced to only 0·7 tonne/ha.

Large differences in erosion thus result from relatively small changes in plant density, and there is a simple explanation for this. It is usual to talk about 'increasing the cover' because this is easy to visualize, but the erosion actually comes from the bare soil, so what really matters is how little of the soil is uncovered and exposed to the rain. And small increases in the plant density have a big effect on the amount of bare soil, as shown by the following example. At a certain plant density 60 per cent of the ground is covered, leaving 40 per cent exposed. If the plant density is increased by a half, the ground covered becomes 90 per cent, leaving 10 per cent exposed. The cover only increases by a half, but the exposed soil is reduced from 40 per cent to 10 per cent, and so the erosion is reduced to one-quarter.

The essence of crop management for erosion control is therefore to minimize splash erosion by providing maximum protective vegetative cover. As we have just seen, one effective method for row crops is to use high planting rates, and another is to ensure large vigorous healthy plants by the right combination of soil moisture, fertilizer, organic matter, and so on. Even the best crop cannot provide cover during the period before planting and after harvesting, and for protection at these times the return of crop residues as a surface mulch greatly reduces the soil loss. In fact experiments have shown that it is not necessary for the crop residues to be entirely on the surface. They were found to be just as effective when ploughed into the surface layer. This is called *trash farming*, as opposed to *surface mulching*, which is keeping it on the surface.

Rotations can also be effective in maintaining the soil in a non-erodible condition. What is required is a well-aggregated soil (that is, one with a good crumb structure) with plenty of organic matter, especially the coarse fibrous undecomposed kind. Ploughing-in a crop of grass or a leguminous forage crop provides just this condition, and so these crops are useful in rotation with cash crops.

8.3. Mechanical erosion control

8.3.1. Purpose and description

Mechanical conservation works are expensive, they make farming more complicated, and they need regular maintenance. They should therefore only be used when the erosion cannot be controlled without them. They are mainly used on arable land, and this is because arable land is (1) more valuable and (2) more vulnerable. We do not often find mechanical protection on rough grazing land because the cost of construction is high compared with the value of the land. Arable land is usually sufficiently valuable to justify the expense. Because arable land does not have permanent vegetation to protect it, like grassland or forest, it is more vulnerable and in greater need of protection.

The purpose of all mechanical conservation is to control the surface run-off. Without any control it flows downhill, increasing in speed and amount as it goes. The object is to stop this happening by keeping it in small manageable flows which can be safely led off the land to the streams and rivers. There are many devices to do this.

The three basic components of run-off control are shown in Figure 8.3. At the top is the stormwater diversion drain to intercept the storm run-off which would otherwise flow down from higher ground on to the arable land which it protects. It is an open drain, usually in bare earth, and on a gentle gradient. It is the first line of defence, and all the structures lower down will be designed on the assumption that it will effectively control all the run-off from outside the arable land. If it fails to do this, the water released will almost certainly breach the lower works. It may be called a *storm drain, stormwater channel, diversion terrace,* or *diversion ditch.*

The run-off from the arable land is caught in similar but smaller drains spaced at regular intervals down the slope. The channel is usually kept free from vegetation, and the excavated soil forms a bank on the downhill side. They are usually on a gentle gradient to lead the run-off safely off the arable land. In different countries they may be called *channel terraces,* or *graded terraces,* or *contour ridges.*

The stormwater drains and the channel terraces can discharge into a natural drainage channel if there is a suitable one in a convenient position. When this is not possible, an artificial channel must be provided. This will also be a shallow open drain, but with a good grass cover, and running straight down the slope. It needs careful design to avoid the flow causing erosion. The usual names are *grass waterway* or *meadow strip.*

Before discussing the design of these three main control measures some others should be described briefly.

Contour cultivation. On gentle slopes it may be sufficient to slow the surface run-off by carrying out tillage operations on the contour. This

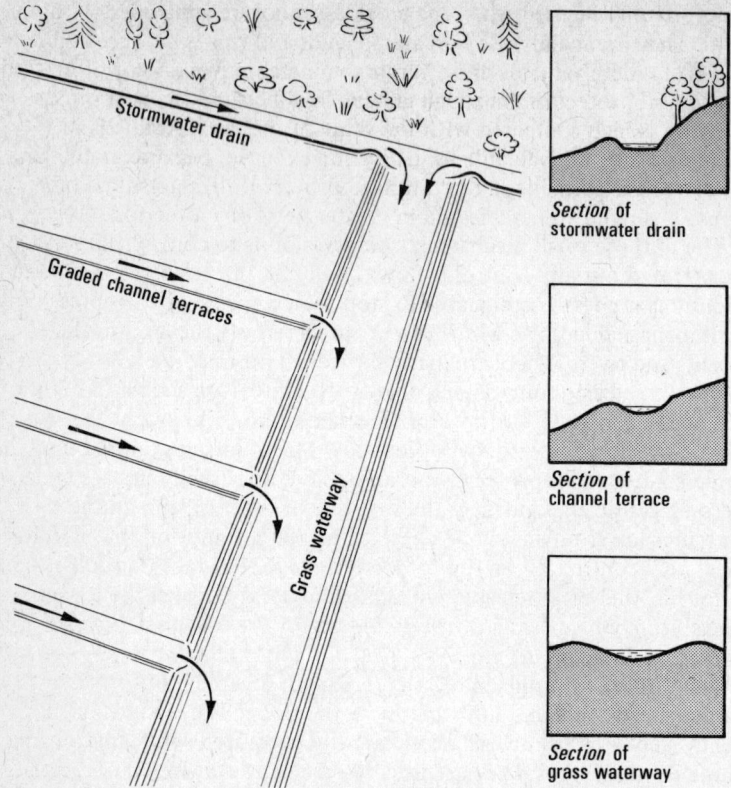

Figure 8.3.   The basic components of mechanical protection: (a) the stormwater drain which diverts storm run-off originating off the arable land; (b) the graded channel terraces leading away the run-off from the arable land; (c) the grass waterway into which both stormwater drain and channel terrace discharge.

is called contour cultivation. It avoids the problem of leaving downhill furrows which concentrate the run-off and cause scour erosion.

Grass strips. Grass strips may also be used when the erosion is not severe. Strips of grass or other close-growing vegetation are left between bands of cropped land. Run-off flowing downhill on the cropland is slowed by the grass strips, and silt which is carried in the run-off is deposited.

Ridging and tied-ridging. Ridging is the construction of small parallel ridges on the contour as part of the tillage operation for row crops. Each ridge acts like a miniature channel terrace. They may be on grade to get surface run-off away, or level if the object is to increase infiltration of water into the soil.

Tied-ridging is building smaller ridges at intervals across the furrows, so that the surface is transformed into a series of small enclosed basins, which store the surface water until it can soak into the soil. The system is effective on deep permeable soils, but should not be used on soils which are too shallow or too impermeable to be able to absorb all the rain.

Contour bund. This is the name used, particularly in India, for another combined soil-conservation and water-conservation practice. They are like extra-large channel terraces on level grade with the ends turned up-hill. The storage capacity is normally sufficient to impound all the surface run-off until it infiltrates, but in order to cater for the exceptional storm there are emergency overflows like small dam spillways.

Pasture furrows. Pasture furrows are for water conservation but should be mentioned because they are similar in appearance to the previous soil-conservation works. The object is to make surface water spread out evenly over grassland. Small open drains are excavated on the true contour, without any bank on the downhill side, so that the run-off spills evenly over the whole length of the drain.

Ridge and furrow. This practice is part soil conservation and part surface drainage, and particularly suitable for large areas of gently sloping land, which do not quite justify channel terraces, but need some control of the surface water. The ground is tilled into wide parallel ridges about 10 or 15 m (30 - 40 ft) wide, with intervening furrows about half a metre deep. Surface water flows across the ridge to the furrow and then down the furrow which is on a gradient of about 1 in 400.

Parallel channel terraces. Conventional channel terraces are laid out on a fixed gradient, and if the land slope varies, so does the distance between adjacent terraces. Cultivating strips of land of varying width is inconvenient, and to avoid this, parallel terraces are sometimes used (Figure 8.4). A *keyline* is chosen, representing a typical terrace line for that part of the field, and channel terraces are laid out parallel to it on either side. When the land slope changes, a new keyline starts another group of terraces. A wedge-shaped piece is left between groups of

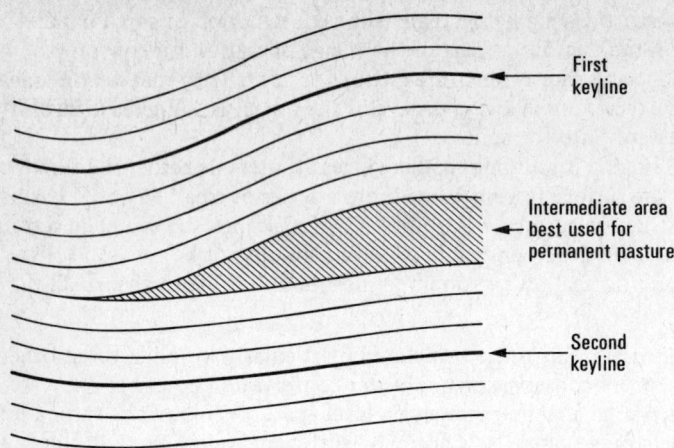

First
keyline

Intermediate area
best used for
permanent pasture

Second
keyline

Figure 8.4.   The 'keyline' method of laying out parallel channel terraces.

terraces and is left uncultivated. The design, layout, and construction of parallel systems is more complicated than conventional systems but may be worthwhile in order to simplify tillage operations.

Bench terraces. Bench-terracing is converting a steep slope into a series of steps with horizontal ledges and vertical or steeply sloping walls between the ledges. If the soil is very stable it may not be necessary to support the wall, but if it is less stable, supporting walls may be built using stone or occasionally brick or timber. A great deal of effort has to be put into constructing bench terraces so their use is limited to three situations:

(1)   when there is an abundant supply of cheap labour, or
(2)   when they will be used for production of a high-value crop such as fruit or vegetables which can justify high construction costs, or
(3)   when the population pressure on the land makes it necessary to use every hectare for food production.

Irrigation terraces. Sometimes bench terraces are built with a raised lip at the outer edge so that they can be used for flood irrigation, especially of rice. This type of terrace is found in most regions where rice is the staple food, and is very common throughout south-east Asia. For rice, the terraces are also level along their length, so that each becomes a flooded shallow pond, but where intermittent water application is intended there can be a longitudinal fall similar to that for border irrigation plots.

Step terraces. Tree crops are often grown on small horizontal terraces just wide enough for one row of trees. For low-growing crops like tea and coffee, each terrace may be from 1 m to 1·5 m (3 - 5 ft) wide and the terraces are continuous over the whole hillside. For larger trees which require wider spacing, such as rubber, coconut, or oil palm, the terraces are small, and spaced at intervals with a cover crop grown on the slope between the terraces. When the terraces are continuous they are called step terraces or orchard terraces. If the levelling is only done for a short distance at each tree-planting station the unconnected terraces are called platforms.

Hillside ditches. These ditches are a kind of channel sometimes built on steep slopes with the emphasis on digging a deep ditch to catch the soil and water. The spoil is used to build a bank on the downhill side of the ditch. There is no evidence that the deep ditch has any merit compared with the shallower channel terrace.

8.3.2. Design and construction

Design life. Protection works need to be designed and constructed carefully. If the drain channel is too small the run-off will overtop and breach the bank. This usually causes serious damage because the water discharges downhill in a concentrated and fast-moving stream. On the other hand, to make the drains big enough to carry the size of flood which has a 100-year probability would be unreasonably expensive. The usual practice is to design farm-scale conservation works for a 10-year flood.

Maintenance. Care in designing the works must be followed by equally careful construction, and by regular maintenance. Earthworks are easily damaged, and planned regular inspection is required so that corrective action can be taken early. A major fault with many government conservation schemes is that no provision is made for subsequent maintenance.

Planning. Conservation measures should be part of an integrated plan of development. The planner must ask questions like:
    'Will the solution of today's problem interfere with future developments?'

'Will the drains and waterways affect access to a neighbouring village?'
'Does the road system fit in well with the drains?'

Mechanical conservation works will interact with all other aspects of land use and management, so questions like these must be considered when planning the conservation works.

Design. Flow in open channels, and the general principles of channel design, were discussed in Chapter 4. The various kinds of conservation works all need a designed channel, but each kind is used only within a certain range of conditions, so the design of each kind can be a simplified limited solution of the general case.

Stormwater drains. A good design method for stormwater drains is the *Durbach method*, which, while based on sound engineering principles, breaks down the operation to a number of small steps, each of which does not require anything more than simple arithmetic.

Table 8.1
Maximum safe velocities in stormwater drains

Material	Maximum velocity on cover expected after two seasons					
	Bare		Medium grass cover		Very good grass cover	
	m/s	ft/s	m/s	ft/s	m/s	ft/s
Very light silty sand	0·3	1·0	0·75	2·5	1·5	4·5
Light loose sand	0·5	1·5	0·9	3·0	1·5	5·0
Coarse sand	0·75	2·5	1·25	4·0	1·7	5·5
Sandy soil	0·75	2·5	1·5	4·5	2·0	6·5
Firm clay loam	1·0	3·5	1·7	5·5	2·3	7·5
Stiff clay or stiff gravelly soil	1·5	4·5	1·8	6·0	2·5	8·0
Coarse gravels	1·5	5·0	1·8	6·0	unlikely to form very good grass cover	
Shale, hardpan, soft rock, etc.	1·8	6·0	2·1	7·0		
Hard cemented conglomerates	2·5	8·0	-	-	-	-

From Dept. Conservation and Extension, Government of Rhodesia,
Dept. In-service Manual.

Table 8.2
Channel factor X

Percentage slope	V S	0·25	0·5	0·75	1·0	1·5	2·0	2·5
1·0	100	-	16	24	32	48	65	80
0·66	150	-	20	30	40	59	79	98
0·50	200	-	23	34	45	68	90	102
0·40	250	13	26	39	51	77	103	-
0·33	300	14	28	42	56	84	112	-
0·25	400	16	32	48	64	97	-	-
0·20	500	18	36	54	72	109	-	-
0·17	600	20	40	59	78	-	-	-

From Dept. of Conservation and Extension, Government of Rhodesia, Dept. In-service Manual.

Missing figures suggest values are unsuitable.

V is velocity of flow in metres per second (from Table 8.1) and 1 in S is gradient of channel.

The procedure for the Durbach method is as follows:

1. From Table 8.1 select the maximum safe velocity. This is the fastest flow which will not cause scouring of the channel. The design is based on the velocity which can be withstood by the vegetative cover after two seasons, because a small amount of scour during this period is preferable to over-designing the channel for the rest of its life. If it is the intention to encourage a good grass cover, the permissible velocity is higher than for the same soil without vegetation.

2. In Table 8.2 choose a probable gradient (1 in 200, or 1 in 250 are most commonly used), and using this and the velocity from Table 8.1 find the value of X.

3. In Table 8.3 from X, find D.

Table 8.3
Values of D for values of X

X	38	50	60	69	78	86	95	103	110
D (m)	0·2	0·3	0·4	0·5	0·6	0·7	0·8	0·9	1·0

X is the channel factor from Table 8.2 and D is the depth of cut channel in metres.

Table 8.4
Values of F

S D (m)	100	150	200	250	300	400	500	600
0·2	0·35	0·25	0·20	0·20	0·20	0·15	0·15	0·15
0·3	0·60	0·45	0·40	0·35	0·30	0·25	0·25	0·25
0·4	0·90	0·75	0·65	0·60	0·55	0·45	0·40	0·35
0·5	1·30	1·15	0·95	0·85	0·80	0·65	0·60	0·55
0·6	1·80	1·55	1·30	1·20	1·10	0·95	0·80	0·75
0·7	2·25	2·00	1·70	1·50	1·35	1·20	1·05	1·00
0·8	2·80	2·45	2·15	1·90	1·70	1·50	1·30	1·25
0·9	3·40	3·00	2·65	2·35	2·10	1·80	1·60	1·50
1·0	4·05	3·60	3·15	2·75	2·50	2·15	1·90	1·85

From Dept. of Conservation and Extension, Government of Rhodesia,
Dept. In-service Manual.

F is the discharge in cubic metres per second per metre width of channel
for various values of D and S where D is the depth of cut channel in metres
(from Table 8.3) and 1 in S is the channel gradient.

4. In Table 8.4 from the chosen gradient and D, find the value of F.
This is the discharge in cubic metres per second per metre width of
channel.

5. Divide F into the maximum flow in cubic metres per second
which the channel will have to carry. This gives the top width for a
shallow channel of parabolic section. For rectangular sections this width
is reduced by one-third.

Example.

Design a channel to carry 2·5 m³/s on a gradient of 1 in 250 through
a sandy soil. Low fertility of the soil and poor rainfall suggest that the
grass cover after 2 years will only be of medium density.

1. From Table 8.1, $V = 1·5$ m/s.
2. Table 8.2, at $S = 250$ and $V = 1·5$, gives $X = 77$.
3. From Table 8.3, for $X = 77$, depth $D = 0·7$ m.
4. Table 8.4, for $S = 250$ and $D = 0·7$, gives $F = 1·5$ m³/s per metre
width.
5. Width $= Q/F = 2·5/1·5 = 1·7$ m approximately.

The channel design is 1·7 m wide and 0·7 m deep.

As in all problems of channel design there is not a single unique solution, but many alternatives. The value of D obtained in Table 8.3 is the maximum value. Any lesser value may be used for steps 4 and 5 and will give a wider and shallower drain. A greater value of D cannot be used because the velocity would then exceed the permissible limit.

Channel terraces. It is not necessary to design channel terraces individually. Instead, we have general specifications covering each of the four main variables.

1. *Spacing.* The distance between adjacent channel terraces is usually expressed as the *vertical interval*, that is, the difference in elevation. On steep slopes the risk of erosion is greater, so the terraces are placed closer together. A useful formula is

$$\text{vertical interval} = \frac{S + F}{6},$$

where S is the percentage slope, that is, the vertical rise per 100 m horizontally and

F is a factor which varies between 3 for light sandy soils and 6 for erosion-resistant clays or clay loams.

2. *Maximum length.* If a channel terrace is too long the volume of run-off in the channel will become too much and the channel will start to scour. From experience it is possible to specify maximum lengths as shown in Table 8.5. Note that these are maximum lengths in the direction of flow. A channel which sheds run-off on both sides from a

Table 8.5
Maximum lengths of channel terraces

	Sandy soils		Clay soils	
	m	ft	m	ft
Normal maximum	250	900	400	1200
Absolute maximum (if spacing is reduced)	400	1200	450	1500

These distances all refer to the maximum length of flow. A channel which sheds run-off on both sides from a high point can have the maximum length on either side of the high point.

From Dept. of Conservation and Extension, Government of Rhodesia, Dept. In-service Manual.

high point can have the maximum length on either side of the high point.

3. *Gradient.* At one time it was fashionable to construct channel terraces with an increasing gradient so that they could carry an increasing flow as more land contributed run-off. Nowadays a constant gradient is preferred. 1 in 250 is best for terraces with clean-cultivated channels.

4. *Cross-sectional area.* The cross-sectional area is usually less precisely defined than the cross-section of other channels and drains. This is because, first, they are usually built with farm machinery, and close control is not possible, and secondly because the cross-section will change when tillage and cultivation operations are carried out.

A specification used successfully in Africa is to define three minimum dimensions:

(a)   a channel width of 2 m (7 ft);

(b)   a minimum excavated channel depth of 0·25 m (9 in);

(c)   a bank height of 0·25 m (9 in) above original ground level.

Given the design conditions for spacing and length previously discussed this should cater for the run-off based on a 10-year return period.

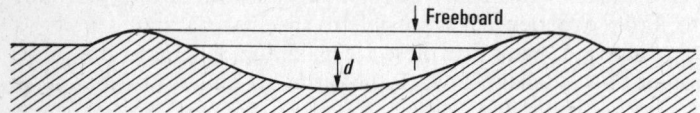

Figure 8.5.   Cross-section of a grass-lined artificial waterway.

Artificial waterways. These need to be individually designed, but the design is simplified because (1) they are nearly always the same shape— that is, a shallow dish-shaped section as in Figure 8.5—and (2) they have a good grass cover, so a constant value can be assumed for the roughness coefficient (Mannings $n = 0·04$). The design procedure for shallow grass-lined channels given in Section 4.2 is suitable for artificial waterways. It is usual to add from 75 mm to 150 mm (3 - 6 in) of freeboard to the design depth as a safety factor.

Construction. The construction of mechanical protection works may be undertaken by the farmers themselves, by private contractors, or by the Government. The advantages and disadvantages of each were discussed in relation to the construction of earth dams in Section 5.1. In general, the same arguments apply to the construction of conservation works, but construction by the farmer is definitely preferable because maintenance is likely to be better.

8.4. Gully control

8.4.1. Causes of gully erosion

Gully erosion is spectacular, but fortunately the kind of land where it is most frequent is usually not very important agriculturally. The control of gullies is always expensive and difficult, and it is very much a case of prevention being better than cure.

Gullies are caused by either of two situations. Either some change leads to an increase in the flood flow, and the watercourse cannot cope with the increase and starts gullying. Or, without the flow increasing, the watercourse may change so that it can no longer cope, and again gully formation results. A common example of change leading to greater floods is a change in the land use, such as converting woodland to arable land. This results in an increase in the maximum rate of run-off, and the previously stable stream starts eroding. An example of the second cause is overgrazing the watercourse. With a good cover of grass or other plants in the bed and on the sides of a stream the water is slowed. After overgrazing there is less resistance to the water which flows faster and starts scouring, especially if the banks have been weakened by the trampling of the stock.

8.4.2. Control methods

Economics. In terms of simple economics, the repair of gullies is seldom justified since the cost of repair is likely to be higher than the value of the land after it has been reclaimed. However, other factors have to be considered before deciding whether control work is justified. Apart from the on-site damage it may be necessary to do something about the gully because of downstream effects, such as a storage dam being silted up or irrigation works being threatened. Or there may be on-site damage other than the obvious loss of land. Examples are the lowering of the water table, damage to fences or roads, or perhaps the possibility that if left unchecked the gully will cut back to a road, bridge, or building. All these factors must be considered before deciding whether to try to control the gully.

Principles. The main principle of gully control is without doubt to determine the cause of the gully and take counter measures. A doctor does not start trying to cure an illness until he knows the nature of the disease and its cause. If the gully was started because the flood flow has doubled as a result of changed land-use, then minor patching up of the damage is not going to solve the problem. If it is not possible to restore the original conditions, then the object must be to create a new channel which can remain stable when carrying the increased flow.

Control by vegetation. Whenever possible vegetation should be used for gully control in preference to structures. Structures of concrete, wood, or any other material will rot and decay and are liable to be undermined or by-passed. Vegetation, however, can multiply and thrive and improve over the years.

The effect of vegetation is to slow down the speed of flow, reducing scour and causing deposition of silt. The deposited soil gives better growth, which traps more silt, and so on until the gully is healed.

The difficulty is that gullies are poor places to establish vegetation. The bed of the gully is probably almost sterile, with no structure, no organic matter, no plant nutrients, and low moisture-holding capacity.

Two ways of overcoming these problems are the selection of suitable plants and the use of special planting techniques. Some useful plants are listed in Table 8,6, but it is difficult to predict which plant will do well, and the best plan is to make trial plantings of a wide range of plants to see which flourish. Local varieties should be investigated. Plants growing reasonably well in or near the gully must be accustomed to the local conditions. Sometimes a little help in the way of fertilizer will enable them to outgrow any imported varieties.

Some techniques to help establishment are described below.

1. Strong seedlings can be established in bottomless polythene cylinders filled with good soil, and planted out in cylindrical holes made with an earth auger. The plastic can be left on. By the time the plant has outgrown its reservoir of fertile soil it is strong enough to survive in the tougher conditions outside. This method is good for establishing *Phragmytes* reed in gully bottoms.

2. To establish colonies of grass, plant into sacks of good soil. The sacks are laid in shallow trenches so they are about level with the gully bed. A small cut is made in the bag and a seedling planted through the cut. The bag prevents the soil and plant being swept away by the first flood, and by the time the bag rots away the plants are established. Old jute or hessian sacks are best, but discarded plastic fertilizer bags or strong paper sacks can also be used.

3. Planting the sides of gullies is difficult because they are steep and unstable. When the cost is justified the banks can be levelled to a gentle

Table 8.6
Some plants useful for gully control

Common name	Botanical name
Star grass (Africa) Bermuda grass (America)	*Cynodon dactylon*
Centipede grass	*Eremochloa ophiuroides*
Weeping love grass	*Eragrostis curvula*
Swaziland finger grass (Africa)	*Digitaria swazilandensis*
Kikuyu grass (Africa)	*Pennisetum clandestinum*
Canary grass	*Phalaris canariensis*
Wheatgrass	*Agropyron* spp.
Kudzu vine	*Pueraria thunbergiana*
Tropical Kudzu	*Pueraria phaseolides*
Taiwan Kudzu	*Pueraria tonkinensis*
Reed canary grass	*Phalaris arundinacea*
Common reed	*Phragmytes* spp.
Lespedezah	or *Lespedeza sericea* *Lespedeza juncea*
Grama grasses	*Bouteloua* spp.
Bluestem grasses	*Andropogon* spp.
Saltbush	*Atriplex* spp.
Sand-bar willow	*Salix exigua*

uniform slope by bulldozers and then seeded or planted. Since the banks are almost sure to be infertile subsoil, some extra fertility must be added. One method is to insert pockets of better soil as was described for planting the floor of the gully. Alternatively a layer of topsoil can be spread over the sloping sides. The danger is that the soil may get washed off before grass can be established. Surface mulching with straw or long grass or any crop residues will help to prevent this.

Temporary structures. It frequently happens that the establishment of vegetation is difficult because the newly planted material gets swept away, or because there is no soil for the vegetation to grow in. In either of these cases there may be a place for temporary structures whose purpose is to provide protection for just long enough to give vegetation a start.

1. *Wire bolsters.* A simple but effective method when there is plenty of loose rock available nearby is to build a loose rock-fill dam with the stones anchored in place by wire netting. Galvanized wire netting of a fairly stout gauge and 2 m (6 ft) or more in width is laid out flat across the gully bed. Loose rock is packed on one-half of the width of the netting and the other half is wrapped over the stones and laced to the other edge.

2. *Netting dams.* Another use of wire netting is to form small check dams, usually near the top end of gullies. Wooden posts are driven into the bed of the gully, and used to support a strip of wire netting which forms a low wall across the gully. The height should be only 0·5 m (18 in) or so and the lower edge of the netting is buried. Light brush or straw is piled loosely against the upstream side of the netting wall and is pressed by the flow of water against the netting to form a barrier which is porous but slows down the flow and causes a build-up of sediment on the upstream side.

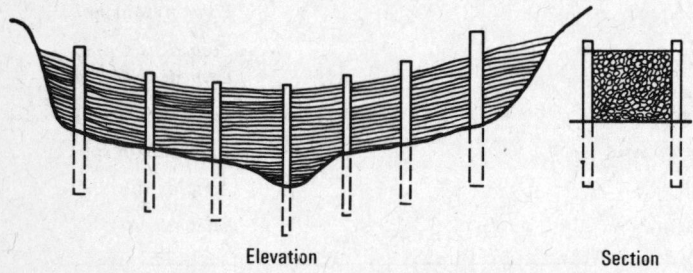

Elevation   Section

Figure 8.6.   A brushwood dam for gully control.

3. *Brushwood dams.* In wooded areas two types of silt-retaining dam are used. The brushwood dam, shown in Figure 8.6, uses small branches up to 25 or 50 mm (1 - 2 in) in diameter, packed as tightly as possible across the direction of flow between rows of vertical stakes. The main points in building these dams are to pack the brushwood as tightly as possible and to secure it firmly. With attention given to both these points it is not uncommon for brushwood dams to last for several years.

4. *Log dams.* When heavier timber is available it can be used for log-piling dams. One method is to use logs in the same way as the brushwood dam but to make a much more substantial structure. Two rows of vertical posts are driven into the bed of the gully and extending up the

sides to above flood level, and then logs are packed in between. The vertical posts should be at least 100 mm diameter, 2 m long, and spaced about 1 m apart in each row, with the two rows of posts 0·5 m apart. In a wide shallow stream it is best to drive in all the vertical posts to about 0·5 m above ground, so that the top of the dam follows the section of the stream bed. If the gully has steep sides it is better to have a rectangular notch in the centre, but the notch must be big enough to pass the whole of the flood.

The second method is to make a simpler structure, consisting only of a single row of vertical posts driven in side by side to form a wall of logs. Again they can either follow the profile of the gully section or have a central notch (Figure 8.7).

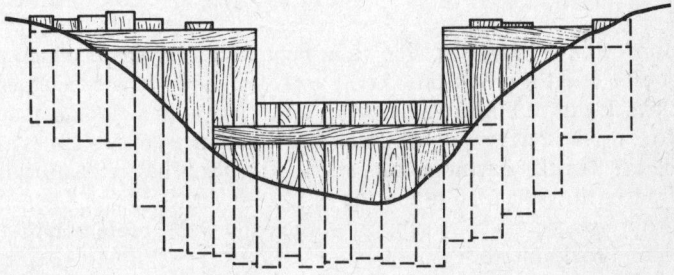

Figure 8.7.   Timber piling used to make a log dam for gully control.

5. *Brick weirs.* Sometimes long-lasting materials are used in structures with a short design life, for example, brick weirs designed to hold sediment long enough to establish vegetation.

Since the object is cheapness and simplicity, the materials and the design must be chosen to suit the site conditions. If the gully bed contains clean washed sand this can be used to make sand-cement bricks very cheaply.

Where clay bricks are the traditional building material they are probably cheap enough for use in gully-control structures. Bricks burned in a kiln are best because they are resistant to the effect of water, but sun-dried bricks will stand occasional wetting and may be used if the gully only experiences infrequent floods. Second-grade bricks and rejects which are not good enough for building but quite adequate for gully control can often be obtained very cheaply from commercial brickfields.

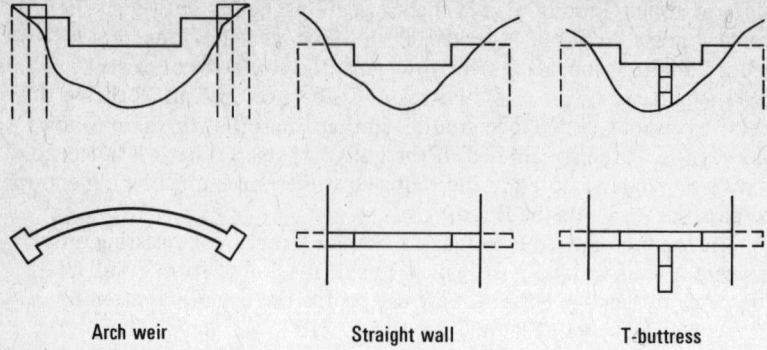

Arch weir   Straight wall   T-buttress

Figure 8.8.   Types of small brick weirs for gully control.

Some simple designs are shown in Figure 8.8. The shape which gives the best strength/weight ratio is the arch weir, and a single thickness of brickwork can be built to a height of up to 2 m with a circular span of up to 2 m. A straight wall of similar size would need 3 or 4 times as much brickwork to achieve comparable strength. The arch wall works by transmitting the load round the arch to the buttresses at each end, and so it needs good solid support in the gully walls, preferably in the form of a rock outcrop.

In the more usual situation of a rock bar which runs across the bed of the gully, a straight gravity-section wall is indicated. The width at the base should be approximately equal to the maximum height, and successive courses of brickwork are narrower so that the section is roughly triangular. Extra strength can be provided by one or more buttresses on the downstream side. The construction of weirs was also discussed in Section 4.2.

Permanent structures. Wherever possible gully control should be achieved by vegetative methods or a combination of vegetation and cheap simple structures whose life is not important. However, there are cases where the problem can only be solved by the construction of permanent structures, and for such works to be successful they must be done thoroughly and carefully. Everything is against their being successful. They will be built in adverse conditions, in poor unstable soils, in remote inaccessible areas where maintenance will be poor, and then they will be expected to withstand the onslaught of torrential floods and to last for ever. The gullies of the world are littered with the remains of ruined structures which only demonstrate that half-measures, or jobs done on the cheap, are a waste of time and effort in gully control.

1. *Silt-trap dams.* One example of where the problem can best be solved by a permanent structure is the case of an excessive sediment load which threatens downstream water supplies. Trapping the silt in sufficient quantity by vegetative means may be slow and uncertain. A quick positive reduction in sediment movement can be achieved by building permanent silt-trap dams. The requirements and design of such dams are the same as for the water-storage dams described in Chapter 5. The object is maximum storage capacity for minimum cost, whether the storage is for water or silt. As for water, storage in a few large dams is usually more economic than in a larger number of smaller dams.

2. *Regulating dams.* Another useful application of permanent dams is to regulate flash floods by what is sometimes called the 'leaky bathtub' principle. A permanent dam is built at the top of the gully with sufficient storage for the run-off from a single storm.

The outlet consists of a permanently open pipe of about 150 - 200 mm (6 - 9 in) diameter which allows the flood water to drain away in a day or two, leaving the storage reservoir empty for the next storm. The flow down the gully is now reduced to the flow through the outlet pipe, so it is fairly easy to create stable conditions which can cope with this flow. Practical construction points are to have the inlet to the pipe raised above the bottom of the dam so that it is not in danger of being silted up, and to discharge the pipe outlet into an energy-dissipating chamber, not directly into the bed of the gully.

3. *Gully-head dams.* A third example of a case for permanent structures is when an active gully head is eating its way steadily upstream, and must be stopped before it threatens a road or bridge or similar asset. An effective way of controlling the erosive force of the run-off over the gully head is to submerge the head of the gully in the pond of a permanently impounding dam. The energy of the in-rushing water is then dissipated as it flows into the pond.

4. *Drop structures.* The other approach to this problem is to stabilize the head of the gully with a masonry, brick, or concrete structure which allows the flood run-off to pass over harmlessly. A typical concrete drop structure is shown diagramatically in Figure 8.9. It is possible to increase the capacity of such a structure without increasing the over-all width by adding a box-inlet on the upstream side, as in Figure 8.10.

To prevent the structure being undermined or by-passed the following points should be observed (see Figure 8.9).

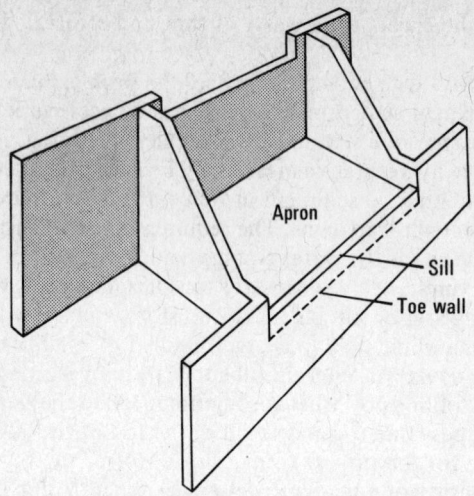

Figure 8.9.   A typical concrete drop structure for gully control.

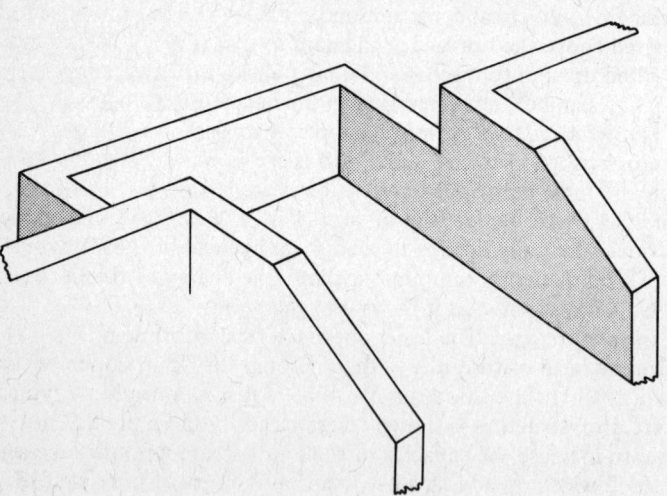

Figure 8.10.   The box-drop inlet increases the capacity of a structure.

1. The notch must be big enough to take the biggest probable flood (a 10-year or 25-year return period is usually used).
2. The wall should be keyed into the banks for a distance equal to the height of the wall.
3. A strong apron is required to withstand the force of the falling water. A sill built at the downstream edge of the apron will form a pool of water, called a stilling basin, which cushions the impact. In erodible soils a toe wall is also required at the downstream edge of the apron to prevent undercutting.
4. Some protection of the sides is usually necessary to prevent erosion of the gully sides by the turbulent water leaving the structure.
5. *Gabions.* The main difficulty with rigid structures is that they cannot adapt when changes occur in the soil surrounding them or supporting them. Concrete, masonry, and brickwork all have good resistance to compression but fail easily under the tension forces which result from settlement. A construction method which overcomes this problem is a more sophisticated version of the wire-netting bolsters described among the temporary measures. The method was developed in Italy and uses prefabricated rectangular baskets, called *gabions*, made of heavy-duty wire netting. The basket is placed in position and filled with stones, then the lid is wired down. The baskets, 4 m long by 1 m by 1 m, are built up on top of each other like courses of brickwork, and can form large or small structures. The use of galvanized wire ensures a long-lasting resistance to corrosion. The main advantage of these structures is that if scouring of the foundations takes place there is sufficient flexibility for the structure to settle slightly without any loss of strength.

8.5. Wind-erosion control

8.5.1. Land and crop management

The first principle of wind-erosion control is the same as for rainfall erosion, that is, correct land-use. When the cause of the erosion is unsuitable land-use, no control treatments will be successful without the restoration of an ecologically sound use of the land.

The classic example of wind erosion caused by unsuitable land use was the 'Dust Bowl' in the central plains of America in the 1930s. Because the rainfall is generally low with periodic severe droughts, the natural vegetation is open prairie grassland, but the settlers, coming from the east and accustomed to better rainfall, ploughed the prairie and planted wheat. Sometimes there would be good rains and good crops, but then a period of drought years would bring crop failure and dust storms. The history of the area is recurrent cycles of development, when the plains were ploughed and farmhouses built, followed by them being abandoned a few years later when the next drought struck. The problem was only brought under control by a massive Government programme aimed at getting the land back to its proper use which is permanent grass.

Assuming that the form of agriculture is right for the ecological conditions, much can be done to reduce wind erosion by the choice of crop-management practices, some examples of which will be briefly described.

Soil only blows when it is dry, so anything which conserves soil moisture is beneficial. Partial cropping may help by reducing the evaporation. This can be done by cropping the same ground in alternate years (hoping there will be a build-up of moisture during the following year) or by cropping part of the land in strips, leaving the remainder in fallow. Another approach to moisture conservation is mulching the soil surface, which both increases infiltration and reduces surface evaporation.

Maintaining a protective cover on the surface of the soil is the key to management for wind control just as it is for control of rain erosion. In addition to helping to conserve moisture, it also slows down the wind close to the soil surface.

Rotations and crop management. Much can be achieved by the careful planning of rotations, especially if there are annual periods particularly liable to strong winds.

The inclusion of periods of grass in rotations is beneficial. In addition to the grass cover ensuring good infiltration and restricting soil movement, it also leaves an improved physical structure and increased organic matter for the grain crops which follow.

Harvesting methods which leave the maximum amount of crop residue are now commonplace. A standing stubble is more effective than when it is chopped and spread on the surface, and the taller the stubble the better. Hybrid varieties with very uniform growth make it possible to set combine harvesters high to take very little of the stalk with the heads.

Conventional methods of land preparation and weed control aim to invert the soil and bury the surface vegetation. Since this is just the opposite of what is required for wind-erosion control, special tillage practices and implements are often used. *Chisel ploughs* can be used instead of discs or mouldboards, and subsurface tillage gives weed control without disturbing the surface. Two implements used for this purpose are the *rod-weeder* and the *duck-foot cultivator*.

Another approach to reducing wind erosion is to slow down the wind by physical barriers. These may be in the form of an increased roughness of the soil surface or planted vegetative barriers. Ordinarily in arable farming the objective is to create as fine and smooth a seedbed as possible, but when wind erosion is likely the opposite is the case. The surface should be as rough and cloddy as is compatible with being able to seed the crop. For row crops the roughness may be further increased by ridges about 200 - 300 mm (8 - 12 in) high, spaced about twice this distance apart. The ridges must run at right-angles to the prevailing wind direction.

Vegetative barriers may be created by strip-cropping with alternate bands of tall and short crops. When the land is to be ploughed, leaving strips of stubble unploughed has the same effect—anything which breaks the flat even regularity is helpful.

Windbreaks and shelter belts are the other way of creating physical barriers, and they are efficient and successful in areas with a rainfall of 400 - 500 mm (15 - 20 in). However, in many areas where wind erosion is prevalent, the low rainfall is likely to be insufficient for good tree-growth. This is important because the effect of windbreaks is a function of their height. For example, the wind velocity is reduced to 40 per cent of the open-air velocity at a distance downwind of 4 times the height of the windbreak. Low windbreaks will therefore only be effective if they are very closely spaced, and in that case they will interfere with farming operations.

8.5.2. Tillage methods

For wind-erosion control the objective is to keep the soil as rough as is compatible with growing the crop. In general terms, any implement or operation which roughens is desirable, and any which pulverizes and smooths is undesirable. A convenient way of describing the implements used is therefore under these headings.

Implements which roughen. The main implements in this group are ploughs, ridgers, and cultivators.

1. Disc-ploughs are more popular than mouldboards because they have a lighter draft, and are better in rough or stony conditions, or working through a heavy mulch. They also bury crop residues less than

mouldboards, and this is also desirable. Chisel ploughs are very suitable, either the rigid tine or spring-shank type. The stump-jump chisel plough used in Australia and Africa prevents shock load damage if it strikes boulders or tree stumps.

2. Ridgers, called listers in America, are like ploughs with two opposed mouldboards, the most common size being 360 mm (14 in) across the wings. They are usually mounted on a tool bar so that the spacing between ridges is adjustable.

3. The duck-foot cultivator was mentioned earlier for subsurface weed control, but it is also widely used for seedbed preparation. Many different shapes of shank and of working points are used in different conditions, and the choice of these and the speed of operation will give different effects. The main point is that the action should be bursting through and stirring the soil, roughening it and bringing up clods, not pulverizing or smoothing.

Implements used for particular purposes

1. The one-way disc is something between a disc-plough and a disc-harrow, with large discs up to 600 mm (2 ft) diameter. It is useful in the system called trash farming, when large amounts of crop residue are partly buried in the plough layer.

2. Shallow cultivations by small duck-foot sweeps have been mentioned for seedbed preparation and for weed control, and a similar implement with much larger V-shaped sweeps is used as a primary tillage tool when crop residues are on the surface. The blades are usually from 1 m to 2 m (3 - 7 ft) wide and a rolling coulter cuts residues ahead of the shank. The depth of operation is up to 150 mm (6 in). Like all subsurface tillage implements it can only be operated in soils which are free from stones, boulders, or tree roots.

3. Planting equipment suitable for wind-erosion conditions is as important as tillage equipment. Desirable features are the ability to seed through a surface mulch and to leave the surface rough. A very successful type is the deep-furrow drill which works well in heavy residues, pushing up a ridge on either side and planting in the furrow between.

Undesirable implements. The implements to be avoided in wind-erosion areas are those which break down the soil, and leave it smooth and vulnerable. All harrows fall in this category, with disc-harrows and spike-tooth harrows being the least desirable. Spring-tooth harrows are less damaging, but chisel cultivators will do a similar job with less pulverization. Power-driven rotary cultivators should be avoided, for they give a well-aerated finely-divided seedbed which is ideal for germination but is the condition most vulnerable to wind erosion.

Appendix: Conversion factors

See also the *Note on units* on p. xiii. The international metric system (SI) has been used as much as possible in this book, so millimetres (mm) are used instead of centimetres (cm). However the SI system recognises that because of established usage the use of some non-SI units will continue. In agricultural engineering the most important of these are the hectare (10^4 m^2) and the litre (10^6 mm^3).

British to metric		Metric to British	
Length			
1 in	= 25·4 mm	1 mm	= 0·0394 in
1 ft	= 0·305 m	1 m	= 39·37 in
1 yard	= 0·914 m		= 3·281 ft
1 mile	= 1·609 km		= 1·094 yard
		1 km	= 0·6214 mile
Area			
1 in^2	= 645·2 mm^2	1 mm^2	= 0·00155 in^2
1 ft^2	= 0·0929 m^2	1 cm^2	= 0·155 in^2
1 yd^2	= 0·836 m^2	1 m^2	= 10·764 ft^2
1 acre	= 0·4047 ha		= 1·196 yard^2
1 mile^2	= 2·590 km^2	1 ha	= 2·471 acre
			= 107 640 ft^2
		1 km^2	= 0·3861 mile^2
			= 247·1 ha
Volume			
1 in^3	= 16387·1 mm^3	1 mm^3	= 0·000 061 in^3
	= 16·387 cm^3	1 cm^3	= 0·061 in^3
1 ft^3	= 0·02832 m^3	1 m^3	= 35·314 ft^3
	= 28·32 l		= 1·308 yard^3
	= 6·23 British gall	1 l	= 0·0353 ft^3
	= 7·48 US gall		= 0·220 British gall
1 yard^3	= 0·7646 m^3		= 0·264 US gall
1 Brit gall	= 4·546 l		
1 US gall	= 3·785 l		

1 Brit gall = 1·132 US gall
1 US gall   = 0·833 British gall
1 acre-in   = 3630 ft^3
            = 102·8 m^3
            = 22650 British gall
1 acre-ft   = 1·23 × 10^3 m^3
            = 272 × 10^3 British gall

Weight

The word 'ton' can mean the British ton of 2240 lb, (called the long ton) or the American ton of 2000 lb (called the short ton), or the metric ton of 1000 kg. The metric ton is sometimes spelled tonne and at 2205 lb is nearly the same as the British long ton of 2240 lb.

1 lb   = 453·6 g   1 kg   = 35·274 oz
        = 0·4536 kg   = 2·2046 lb

British to metric	Metric to British
1 British ton (of 2240 lb) = 1016 kg = 1·016 tonne	1 tonne = 2204·6 lb = 0·9842 British ton = 1·1023 US ton
1 US ton (of 2000 lb) = 907·2 kg = 0·9072 tonne	

Rates of flow

British to metric	Metric to British
1 ft^3/s (cusec or cfs) = 0·0283 m^3/s = 28·32 l/s = 101·94 m^3/hour = 374 British gall/min = 449 US gall/min = 1 acre in/hour (approx.)	1 m^3/s (cumec) = 35·315 ft^3/s = 13230 British gall/min = 15852 US gall/min
1 ft^3/min = 1·7 m^3/s = 4·546 l/min = 0·2728 m^3/hour	1 m^3/min = 0·5886 ft^3/s = 220·5 British gall/min = 16·67 l/s
1 US gal/min = 0·0631 l/s = 0·227 m^3/hour	1 m^3/hour = 0·278 l/s = 3·668 British gall/min = 4·40 US gall/min

$$1 \text{ l/s} \quad = 0.035 \text{ ft}^3/\text{s}$$
$$= 13.21 \text{ British gall/min}$$
$$= .792.5 \text{ British gall/hour}$$
$$= 15.85 \text{ US gall/min}$$

Power

1 hp   = 0.746 kW   1 kW   = 1.34 hp

Pressure

The units of pressure need more explanation because SI units are different from both the English and metric systems.

Pressure is defined as force per unit area, and the difference arises from the different ways of defining force. In both the English and metric systems the unit of force is the same as the unit of weight, that is mass multiplied by the acceleration due to gravity. In SI the unit of force is the newton, defined as the force which, when acting on a mass of one kilogram, gives it an acceleration of one metre per second per second. Thus the SI unit of pressure is based on unit acceleration not on gravitational acceleration. To convert from metric pressures we have to multiply by gravitational acceleration which is 9.81 m/s^2. Some short-cut approximations can be made by using 10 instead of 9.81.

For convenience, pressures are also sometimes defined by comparison with atmospheric pressure, but since atmospheric pressure varies, some arbitrary standard must be chosen. The British standard atmosphere is 14.70 lbf/in^2, but it not used very much. The metric standard atmosphere is slightly less (equivalent to 14.50 lbf/in^2) and is used more. It is called the bar, and subdivided into thousandths, or millibars. In SI units the bar is 100 kN/m^2 and the millibar is 100 N/m^2. The hectobar (one hundredth of a bar) is also used.

When precise accuracy is not required, and an error of 2 per cent is acceptable, a simple conversion may be made from metric units to SI by taking 1 kgf/cm^2 as equal to 1 kN/m^2 instead of the true value of 0.981 kN/m^2. This is useful because much practical data on pressures is already listed in kgf/cm^2.

The third way of describing pressure is by the height of a column of liquid which will exert a given pressure. The liquid is either water or mercury (Hg), and both British and metric units of length are used.

British to metric	Metric to British

(a)   Accurate conversions

1 lbf/in²	= 0·0703 kgf/cm²	1 kgf/cm²	= 14·22 lbf/in²
	= 2·31 ft of water		= 32·81 ft of water
	= 0·703 m of water		= 10 m of water
	= 2·04 in of mercury		= 28·96 in of mercury
	= 51·82 mm of mercury		= 735·6 mm of mercury
1 lbf/ft²	= 4·88 kgf/m²	1 kgf/m²	= 0·205 lbf/ft²

(b)   SI conversions

1 lbf/in²	= 6·895 kN/m²	1 kN/m²	= 0·145 lbf/in²
1 lbf/ft²	= 47·9 N/m²	1 N/m²	= 0·021 lbf/ft²
1 kgf/cm²	= 0·981 kN/m²	1 kN/m²	= 1·02 kgf/cm²
1 kgf/mm²	= 9·81 N/m²	1 N/m²	= 10·2 kgf/m²

(c)   Practical approximate conversions

1 bar	= 100 kgf/cm²	= 10 m of water
1 lbf/in²	= 7 kN/m²	= 0·07 kgf/cm²

Glossary

aggregate. In soil science aggregates are distinct particles of soil made up from smaller particles which are joined (aggregated) together. In civil engineering the aggregate is the crushed stone, gravel, etc. mixed with cement to make concrete.

bed rock. Unweathered rock lying below soil and decomposing rock.

berm. A flat ledge, often including a drainage channel, used to break a long slope of an earth embankment.

borrow pit. The area from which soil is taken to build an earth dam or embankment.

consolidation. Making earth more dense by rolling, ramming, compacting with machinery, etc.

consumptive use. The amount of moisture taken from the soil when a crop is grown. It may be expressed as the amount of water required to grow the crop (e.g. 500 mm), or as a rate of usage (e.g. 5 mm per day, or 150 mm per month). See also *evapotranspiration.*

core. A clay core is an impermeable wall of clay to prevent seepage through an embankment. The core trench is the trench cut into the foundations of the embankment.

crest width. The width of the top of a dam wall or weir.

criteria. The tests or standards used to judge whether a specification has been met. In classification systems there is a set of criteria for each class.

cut-off time. In surface irrigation, the time at which the supply of water is cut off at the top end, leaving the water already supplied to flow down the slope.

deflocculation. The break-up or dispersion of soil crumbs or aggregates.

ecology. Ecology is the study of man, plants, and animals in relation to the environment. 'Sound ecology' implies that there is a stable long-term balance between the environment and the activities of man.

erodibility. The lack of resistance of a soil to soil erosion.

erosivity. The ability of rainfall to cause soil erosion.

empirical formula or empirical equation. An equation or formula based on observed facts or experimental results, not derived from theoretical considerations.

evapotranspiration. The total loss of moisture from the soil to atmosphere including both evaporation from the soil surface and plants, and transpiration from the plant leaves. It is usually expressed as a rate of loss (e.g. 5 mm per day). See also consumptive use.

flood routing. Calculating the balance of inflow, storage, and outflow for a stream, river, or reservoir.

freeboard. The extra height of a dam wall above the full supply level of the reservoir, to allow for settlement, wave action, or flood levels.

hydraulic conductivity. The rate at which water moves through soil. It is expressed as m per day.

impeller. The part which in a pump turns a rotary force into a force of propulsion on the water, and which in a turbine turns the water movement into rotary power. A propeller is an impeller with tilted blades like a fan so that the water flow is along the axis of the propeller shaft.

infiltration. The rate at which water enters the soil surface, as opposed to percolation which is the movement through the soil below the surface. It is expressed as mm per hour or in per hour. See also permeability and hydraulic conductivity.

intensity. The rate of rainfall, expressed as mm per hour or in per hour.

laterals. In a sprinkler system, the pipes that the sprinklers are connected on to—so called because they go out laterally, i.e. on either side of the main distribution pipe.

laterite. A soil containing a high proportion of iron or aluminium hydroxide. Often it is 'indurated', i.e. hardened as though it were cemented together.

leaching. The washing-out of any undesirable soluble chemicals by applying excessive amounts of water.

mole drains. The sub-surface drains formed by pulling a cylinder through the soil to leave a circular open drain.

PTO (power take off). A place on a tractor where power from the engine can be taken off to drive auxiliary equipment, whether the tractor is stationary or not. Usually in the form of either a pulley, or a rotary drive-shaft.

permeable fill. A porous material which is used in sub-surface drainage systems to improve the rate of flow of water to tube drains.

permeability of soil. The rate at which a fluid (air or water) moves through soil. See also *hydraulic conductivity.*

probability. An event, such as a flood or a rainstorm, with a 10-year probability is one which *in the long term* will occur with an average frequency of once in 10 years. A 10-year probability may also be referred to as a 10-year return period. The ratio of different return periods is shown in Table 3.2 (p.42). Probability applies only to long-term records and short-term variations can and will occur.

propeller. See *impeller.*

puddling. Making a clay soil impermeable by manipulating it and compacting it while it is wet.

return period. See *probability.*

savannah. Large open areas of land, usually fairly level, with low rainfall, low vegetation, and few or no trees.

side slopes. The steepness of the slope of the sides of an earth wall or a canal cut in earth. It is expressed as the ratio of horizontal to vertical.

soil morphology. The form (i.e. shape, nature, and texture) of soils.

soil pedology. The study of soils, particularly their origin.

spillway. When a dam is full, any further incoming water flows over or through the spillway.

> *natural spillway* —a spillway of undisturbed natural vegetation
> *cut spillway* —an open channel cut in soil or rock to give the required cross-section and gradient
> *grassed spillway* —a cut spillway planted with grass
> *mechanical spillway* —a tube or pipe through the dam, or a concrete or masonry channel or chute.

stone pitching. Large stones or boulders individually placed and tightly packed, to give protection against scour erosion on the bottom or sides of an earth channel, or on a sloping earth bank. Called rip-rap in the U.S.A.

tile drains. Tube drains made from burnt clay, originally in the same way as roofing tiles. Sometimes extended by analogy to tube drains made from other materials especially concrete.

topographic map. A map which shows all the physical features of an area, such as hills, rivers, roads, and towns.

training wall. A low earth wall built as an extension of the main embankment of an earth dam. The purpose is to divert away from the main embankment flood water which is passing through a cut or natural spillway.

tranquil flow. Flow in a stream or canal which is smooth and even. The opposite of turbulent flow.

trajectory. The curved path of a jet of water through the air after it leaves a pipe under pressure.

turbulent flow. Flow in a stream or canal when there are swirls and eddies. The opposite of tranquil flow.

waterlogged soil. Soil which is unmanageable because of its excessive wetness.

Further reading

BEERS, W.F.J. VAN (1958). *The auger-hole method.* Bull. 1, International Institute for Land Reclamation and Improvement, Wageningen, Netherlands.

HUDSON, N.W. (1973). *Soil conservation.* Batsford, London, and Cornell University Press, New York.

HUDSON, A.W., HOPEWELL, H.G., BOWLER, D.G. and CROSS, M.W. (1962). *The draining of farm lands.* Massey Agricultural College, New Zealand.

PEREIRA, H.C. (1973). *Land use and water resources.* Cambridge University Press.

SCHWAB, G.O., FREVERT, R.K., EDMINSTER, T.W., and BARNES, K.K. (1966). *Soil and water conservation engineering* (2nd edn.). Wiley, New York.

UNITED STATES BUREAU OF RECLAMATION (1965). *Design of small dams.*

— (1967). *Water measurement manual.*

UNITED STATES DEPARTMENT OF AGRICULTURE (1962). *Soil survey manual.* Agricultural Handbook 18.

— (1973). Soil Conservation Service. *Drainage of agricultural land.* Water Information Center, Washington.

WARD, R.C. (1967). *Principles of hydrology.* McGraw-Hill, Maidenhead.

WITHERS, W.B.J. and VIPOND, S.J. (1974). *Irrigation: design and practice.* Batsford, London.

Index

venturi meter, 60, 62

Washington flumes, 56, 59-60
water
   requirement of crops, 119
   yield, 44
weirs, 109
   brick, 203-4
   concrete, 111

   gravity, 110
   rectangular, 53-5
   90° V-notch, 52-4
wild flooding, 149-50
wilting point, 6, 124
windbreaks, 209
wind erosion, 207
wire bolsters, 202

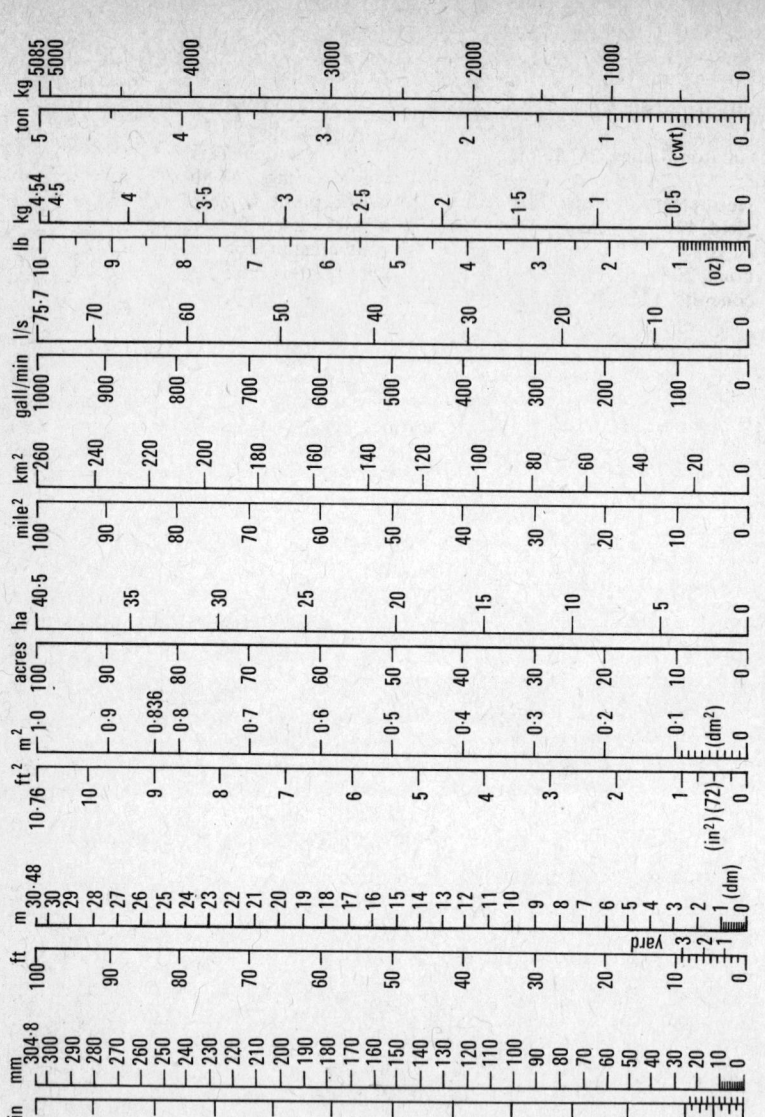

See the Appendix for more detailed conversion factors

ton	kg
5	5085
	5000
4	4000
3	3000
2	2000
1	1000
(cwt)	0

lb	kg
10	4·54
	4·5
9	4
8	3·5
7	3
6	2·5
5	2
4	1·5
3	1
2	0·5
1	0
(oz)	

gall/min	l/s
1000	75·7
900	70
800	60
700	50
600	40
500	30
400	20
300	
200	10
100	
0	0

mile²	km²
100	260
90	240
80	220
	200
70	180
60	160
50	140
40	120
30	100
	80
20	60
10	40
	20
0	0

acres	ha
100	40·5
90	35
80	
70	30
60	25
50	20
40	15
30	
20	10
10	5
0	0

ft²	m²
10·76	1·0
10	0·9
9	0·836 0·8
8	0·7
7	
6	0·6
5	0·5
4	0·4
3	0·3
2	0·2
1	0·1
(in²)(72)	(dm²)
0	0

ft	m
100	30·48
	30
	29
	28
90	27
	26
	25
80	24
	23
	22
70	21
	20
	19
60	18
	17
	16
50	15
	14
	13
40	12
	11
	10
30	9
	8
	7
20	6
yard	5
3	4
2	3
10	2
1	1 (dm)
0	0

in	mm
12	304·8
	300
11	290
	280
	270
10	260
	250
	240
9	230
	220
	210
8	200
	190
7	180
	170
	160
6	150
	140
	130
5	120
	110
	100
4	90
	80
3	70
	60
	50
2	40
	30
1	20
	10
in 0	0